Carbon Markets

Gbenga Ibikunle • Andros Gregoriou

Carbon Markets

Microstructure, Pricing and Policy

Gbenga Ibikunle
University of Edinburgh
Edinburgh, UK

Andros Gregoriou
University of Brighton
Brighton, UK

ISBN 978-3-030-10276-0   ISBN 978-3-319-72847-6 (eBook)
https://doi.org/10.1007/978-3-319-72847-6

© The Editor(s) (if applicable) and The Author(s) 2018
Softcover re-print of the Hardcover 1st edition 2018
This work is subject to copyright. All rights are solely and exclusively licensed by the Publisher, whether the whole or part of the material is concerned, specifically the rights of translation, reprinting, reuse of illustrations, recitation, broadcasting, reproduction on microfilms or in any other physical way, and transmission or information storage and retrieval, electronic adaptation, computer software, or by similar or dissimilar methodology now known or hereafter developed.
The use of general descriptive names, registered names, trademarks, service marks, etc. in this publication does not imply, even in the absence of a specific statement, that such names are exempt from the relevant protective laws and regulations and therefore free for general use.
The publisher, the authors and the editors are safe to assume that the advice and information in this book are believed to be true and accurate at the date of publication. Neither the publisher nor the authors or the editors give a warranty, express or implied, with respect to the material contained herein or for any errors or omissions that may have been made. The publisher remains neutral with regard to jurisdictional claims in published maps and institutional affiliations.

Cover Image © Juhani Viitanen / Alamy Stock Photo
Cover design by Tjaša Krivec

Printed on acid-free paper

This Palgrave Macmillan imprint is published by Springer Nature
The registered company is Springer International Publishing AG
The registered company address is: Gewerbestrasse 11, 6330 Cham, Switzerland

To Ewelina, of course.

Acknowledgements

A significant portion of the empirical work on which several chapters of this book are based were conducted with colleagues we have been honoured to work with over the past few years. In this respect, we acknowledge the contributions of Naresh Pandit, Andreas Hoepner and Mark Rhodes to this project, and we are grateful for their support.

We thank colleagues whose constructive and useful comments have helped define the contents of some of the chapters in the book. Specifically, we thank Seth Armitage, Jo Danbolt, George Daskalakis, Ivan Diaz-Rainey, Jerome Healy, Apostolos Kourtis, Peter Moffatt and Bert Scholtens. A very special note of thanks is due to Ivan Diaz-Rainey and Bert Scholtens.

The analyses conducted in this book would not have been possible without the provision of data by the Intercontinental Exchange Group and the London Energy Brokers Association. Therefore, we thank them for their support.

We would also like to thank our Editor at Palgrave Macmillan, Tula Weis, for helping to bring this book to completion. We thank the Editorial Assistants in the Palgrave Macmillan's Professional Finance division, Ruth Noble and Jazmine Robles for their guidance and work on the project. All members of the commissioning, editorial and production teams at Palgrave Macmillan deserve our gratitude for their professionalism and support.

We are particularly indebted to Aimee Dibbens, our commissioning editor when this project was first agreed. Aimee was instrumental in getting this project off the ground from the start.

Finally, we thank our research assistant at the University of Edinburgh, Khaladdin Rzayev, for helping with all the editorial work.

As always, all the remaining errors are our own.

Contents

1	An Introduction to the Book	1
2	Emissions Trading in Europe: Background and Policy	15
3	Price Discovery and Trading After Hours on the ECX	39
4	The Price Impact of Block Emissions Permit Trades	91
5	The Liquidity Effects of Trading Carbon Financial Instruments	129
6	Liquidity and Market Efficiency in Carbon Markets	165
7	The Future	201
Bibliography		209
Index		225

List of Abbreviations

AAR	Average Abnormal Returns
AHT	After-Hours Trading
AMC	After Market Closes
BMO	Before Market Opens
CAR	Cumulative Abnormal Return
CCFX	Chicago Climate Futures Exchange
CCX	Chicago Climate Exchange
CDM	Clean Development Mechanism
CERs	Certified Emission Reduction Units
CFI	Carbon Financial Instrument
CITL	Community Independent Transaction Log
CLOB	Central Limit Order Book
CO_2	Carbon/Carbon Dioxide
EC	European Commission Regulation
ECC	European Commodity Clearing
ECX	European Climate Exchange
EEX	European Energy Exchange
EFP/EFS	Exchange for Physical/Swaps
EOD	End of Day
EPA	Environmental Protection Agency
ERUs	Emission Reduction Units
ETS	Electronic Trading System
EUAA	EU Aviation Allowances

List of Abbreviations

EUAs	European Union Allowances
EU-ETS	EU Emissions Trading Scheme
GHG	Greenhouse Gas
HAC	Heteroscedasticity and Autocorrelation covariance
ICE	Intercontinental Exchange
IET	International Emissions Trading
ISVs	Independent Software Vendors
ITL	International Transaction Log
JI	Joint Implementation
MEC	Market-Efficiency Coefficient
MSR	Market Stability Reserve
MW	Megawatts
NAP	National Allocation Plans
NYSE	New York Stock Exchange
OTC	Over the Counter
PHA	Person Holding Account
RI	Responsible Individual
RSPR	Relative Spread
RTH	Regular Trading Hours
SSM	Saudi Stock Market
tCO_2	Tonne of CO_2
TRS	Trade Registration System
TSPR	Traded Spread
UNFCCC	United Nations Framework Convention on Climate Change
VAR	Vector Autoregressive Model
WPC	Weighted Price Contribution
WPCT	Weighted Price Contribution per Trade

List of Figures

Fig. 2.1	Response of carbon financial instruments to a recession	29
Fig. 3.1	Trading volume and volatility	49
Fig. 3.2	Log median and mean trade size	50
Fig. 3.3	Unbiasedness regressions by intervals	78
Fig. 4.1	Intraday variations in relative bid-ask spread on the ECX	111
Fig. 5.1	Time series of quoted, relative and effective bid-ask spread estimates	157
Fig. 6.1	Market efficiency measured by 15-minute return predictions with 15-minute lagged Euro Order Imbalance	180
Fig. 6.2	Daily average traded spread and relative spread for ECX, 2008–2012	182
Fig. 6.3	Monthly illiquidity/price impact ratios	183

List of Tables

Table 2.1	Phases of the European Union Emissions Trading Scheme (EU-ETS)	21
Table 3.1	Trading summary	48
Table 3.2	Information asymmetry and half-spread by time interval	66
Table 3.3	Weighted price contribution by time intervals	70
Table 3.4	Weighted price contribution per trade by time intervals	75
Table 3.5	Unbiasedness regressions by intervals	80
Table 4.1	Summary statistics for block trades	104
Table 4.2	Determinants of price impact of block trades	106
Table 4.3	Determinants of price impact and block trade sizes (purchases)	115
Table 4.4	Determinants of price impact and block trade sizes (sales)	120
Table 5.1	Table of events and corresponding dates	136
Table 5.2	Descriptive statistics	137
Table 5.3	Average abnormal returns	141
Table 5.4	Short-term trading volume changes around events	143
Table 5.5	Long-term trading volume around events	145
Table 5.6	Short- and long-term liquidity changes on the EEX	154
Table 6.1	Descriptive statistics for 15-minute liquidity and order imbalance proxies	174
Table 6.2	Correlations for 15-minute trading intervals on the ECX	176
Table 6.3	Predictive regressions of 15-minute returns on lagged order imbalance	177

Table 6.4	Distribution of liquid and illiquid days on the ECX (2008–2011)	184
Table 6.5	Predictive regressions of 15-minute returns on lagged $OIB€_t$ and lagged $OIB€_t$ interacted with an illiquidity dummy	185
Table 6.6	Liquidity and market efficiency dynamics: Granger causality analysis	188
Table 6.7	Daily variance ratios for EUA futures contracts across trading periods	191

1

An Introduction to the Book

This book is based on the insights gained from studies led by the authors between 2010 and 2015; a few of the studies are also extensively described in four of the chapters here. The focus of this book sits at the nexus of three interrelated fields of study: environmental policy, market microstructure and environmental financial economics. Specifically, three main issues (liquidity, price discovery and market efficiency) are investigated using data from the two major emissions permit trading venues within the European Union-Emissions Trading Scheme (EU-ETS). These two venues, along with several others in Europe, constitute the largest regional market for emission permits (see Daskalakis et al. 2011 for detailed discussions; Chap. 2 also provides a descriptive analysis of the market).

According to O'Hara (2003), organised markets fulfil two major functions. The first is the provision of liquidity, while the second is the provision of a mechanism for achieving price discovery. These two, although frequently overlooked by symmetric information-based asset pricing models, are vital to pricing assets of any kind, including emission permits, which are traded in carbon markets. Although distinct in certain respects, liquidity and price discovery are inextricably linked, both theoretically and empirically. In this book, we hold the view that a market is

© The Author(s) 2018
G. Ibikunle, A. Gregoriou, *Carbon Markets*,
https://doi.org/10.1007/978-3-319-72847-5_1

only informationally efficient to the extent to which its instruments or assets traded through it reflect all available information. This includes publicly available information and private information already used in trading, that is, if one limits the argument to the price adjustment process once a foundation price exists. If the instruments reflect all available information, such a market can be considered informationally efficient (Fama 1970). This implies that prices should only move based on innovation in beliefs (developed based on new information), and if prices were to move without supporting information arriving in the market, the market can be considered relatively less efficient.

Liquidity plays an important role in effecting the incorporation of new information into asset prices. O'Hara (2003) makes an interesting analogy: consider a market with only sellers and no buyers on a particular day of the week; unless the sellers are willing to wait for the arrival of the buyers who are expected to arrive on a later day during the week, there will be no trades, hence no liquidity. Now, imagine that an unaligned agent decides to buy off instruments from the sellers on their day of arrival and keeps the instruments until the buyers arrive on a later day, then we have trades, and hence, liquidity. Liquidity is thus simply the process of connecting a buyer to a seller in as frictionless a manner as possible. A spread between the selling and buying prices naturally develops as a result of the service provided by the unaligned agent or middleman. The liquidity of this hypothetical market, based on the spread earned by the agent, is the transaction cost that ultimately affects asset pricing (e.g. Amihud 2002; Grossman and Miller 1988).

Information, liquidity and price discovery are therefore key issues in financial markets, including environmental markets such as the EU-ETS, which is used for trading carbon/carbon dioxide (CO_2) emission permits. The smooth functioning of the EU-ETS can therefore not be evidenced without considering its performance based on the major market functions of liquidity, price discovery and related microstructure functions. This implies that the efficiency of the market relies on its liquidity and price discovery process. These issues are therefore interrelated in a well-functioning market. Evidence suggesting relatedness of these issues can be found in the market microstructure literature (e.g. see Hendershott et al. 2011). In a unique market, such as the EU-ETS, where trading arises as a result of governmental policy, it is important to understand how policy informs the evolution of the market microstructure. As this book aims to address the knowledge gap on the microstructure of the EU-ETS, in the following paragraphs, we review the microstructure literature for financial markets price discovery, price impact of block trades and liquidity by linking them together from the transaction costs perspective. We also link market efficiency to liquidity based on the current literature, since this connection affects transaction costs. We then extend the links from these strands of the market microstructure literature to the growing literature on the EU-ETS. We conclude this introductory section by providing a brief introduction of the empirical contributions contained in this book.

The explanation of liquidity given in the foregoing paragraph creates the impression of illiquidity as the premium paid by a buyer of an asset in a buyer-initiated trade or price concession by a seller in a seller-initiated trade. Thus, the larger the spread between the bid and offer prices the larger the cost of trade. The spread is a necessity reflecting the costs borne by the traders and the economic gain for the intermediary, or as normally referred to in the market microstructure literature, the market maker. Therefore, in quote-driven markets, the market maker quotes provide the basis for measuring transaction costs. The costs, however, are not entirely due to the need for immediacy or processing order costs; they arise as a result of inventory and adverse selection/information asymmetry as well (see Glosten and Milgrom 1985). When considering regular-sized trades, the spread is usually the only microstructure impact on prices; this explains why Brennan and Subrahmanyam (1996) measure illiquidity by price impact. Their measure is based on Kyle's (1985) model and is estimated analogously to Hasbrouck (1991a) and Foster and Viswanathan (1993). They initially employ the Lee and Ready (1991) algorithm in classifying trades into signed trades and then simply estimate the slope coefficient of tick-by-tick price adjustments on signed order flow (size). Their analysis is valid based on several earlier theoretical works (e.g. see Easley and O'Hara 1987; Kyle 1985) and at least an empirical study (see Glosten and Harris 1988), which suggest that the liquidity effects of transaction costs (especially asymmetric information) are captured by trade impacts. Liquidity effects thus play a role in the price discovery process. However, block (large) trades can potentially cause price shocks larger than what the spread components would have (see Chiyachantana et al. 2004; Kraus and Stoll 1972); hence, Brennan and Subrahmanyam (1996) control for earlier information based on order size and price changes. Movement in price by liquidity-motivated block trades is usually due to misinterpretation of trading intentions at execution. These trades, in this case, could be misconstrued as information conveying.

Kraus and Stoll's (1972) contribution is among the earliest studies to establish that block trades do induce price impact. They contend that block trades induce short-run liquidity effects for two reasons. The first is due to price compromise suffered because counter-parties are not readily available and second, due to price compromise when instruments are not perfect substitutes for each other, which leads to inefficient trading and hence price impact. In addition, the idea that price concessions are granted in order to execute an order underscores desperation to make a trade happen. This in itself conveys information to the market about the potential value of the order to the counter-parties, that is, the liquidity constraints they face; the order thus becomes information-laden leading to price impact. Holthausen et al. (1990) find evidence of premium payment or price concession for the execution of buyer-initiated block trades. They argue that buyers in a block trade pay premium; the premium is incorporated permanently in the price consequently, while no evidence of concessions is found for block sales. Kraus and Stoll (1972) further hold that price impact is higher for block purchases than sales because concession or an implicit commission paid is usually higher for purchases than sales. This suggests that there is indeed premium paid on block sales. A major contribution from Kraus and Stoll's (1972) pioneering work is that they establish a relationship between block trades and price impact (see also Chan and Lakonishok 1993). According to the early market microstructure literature, large trades could be viewed as being privately held information conveying; hence it is not unusual for large trades (with no underlying privately held information) to move prices. This potentially complicates the price discovery process in the presence of block trades. The price impact of block trades on financial markets therefore remains an important area of market microstructure research within the context of price discovery.

Since the spread (an inverse proxy for liquidity) retains the tag of being a transaction cost for traders, when large enough, it should negatively affect asset returns and value (O'Hara 2003). Suppose the transaction costs attributable to liquidity were reduced, perhaps through the establishment of a transparent trading mechanism. A mechanism where the true values of assets are reflected hence the market maker need not account for adverse selection costs, then we can safely assume an efficient price discovery process. Studies have reported price impacts on account of switches in platform trading mechanisms (e.g. see Amihud et al. 1997). This does not necessarily relate price discovery and transaction cost components of the bid-ask spread although in this book we present evidence that low adverse selection costs are associated with higher price efficiency for liquid instruments. When the instrument is relatively illiquid, the relationship is not sustained (see Chap. 3). This is an important knowledge gap that this book fills with the evidence provided in Chap. 3.

Price discovery entails the absorption of information into asset prices and is thus always affected by the information content of trades. If a trader is uninformed, they run the risk of trading with informed traders, thus making information a risk issue as far as price discovery is concerned. Uninformed traders must then seek compensation for trading under asymmetric information conditions. The riskiness of their assets cannot just simply be diversified away, since they remain uninformed. Holding more instruments simply add on more potential losses, here the capital asset pricing model (CAPM) world does not exist since the conditions are asymmetric (O'Hara 2003). The presence of informed traders in the market is linked with asymmetric information costs and widening spreads, and widening spreads indicate deteriorating liquidity or rising illiquidity (e.g. see Hasbrouck 1991a, b). Price discovery and liquidity are therefore, based on this view, at the minimum indirectly related. The innovation of liquidity and price discovery also suggests that the degree of pricing efficiency in a market is predicated on these two major functions of market. The next three paragraphs further explore this connection.

As market participants require time to incorporate new information into their trading strategies, a market deemed efficient over a daily horizon does not necessarily translate into a market that is efficient at every point during the day (e.g. see Chordia et al. 2008; Epps 1979;

Hillmer and Yu 1979; Patell and Wolfson 1984). Confirmation of this notion is available in the contributions of Cushing and Madhavan (2000) and Chordia et al. (2005), showing that short-run returns can be predicted from order flows. According to Chordia et al. (2008), this predictability diminishes with improving market liquidity and across different tick size regimes on the New York Stock Exchange (NYSE). Chung and Hrazdil (2010a) also confirm the diminishing predictability proposition in a large sample analysis of NASDAQ stocks. These two studies thus provide evidence of strong relations between liquidity and market efficiency through the impact of liquidity on the predictability of returns from order flows.

Previous studies have taken a different path in linking liquidity and market efficiency. These studies examine the connection between liquidity and returns through the demand for premia when transacting in illiquid instruments. Each of the studies provide varying insights. Pástor and Stambaugh (2003) report the cross-sectional relationship between stock returns and liquidity risks. Their results are in line with findings from Datar et al. (1998) and Acharya and Pedersen (2005). Similarly, Amihud's (2002) document evidence supporting the hypothesis that expected market liquidity provides an indication of stock excess return in the time series, implying that the excess return to some extent typifies an illiquidity premium. Chang et al. (2010) also find a consistent narration on the Tokyo Stock Exchange (TSE).

Chordia et al. (2008) make an insightful argument for the relatedness of market efficiency, price discovery and market liquidity. Consider market makers in a hypothetical market struggling to sustain liquidity supply. This may be as a result of financial difficulties or over-exposure to untenable positions. In any case, when such a scenario exists, pricing strain caused by arriving order flows potentially forces a brief deviation of prices from their underlying worth (hence inefficiency; see Fama 1970). Thus, order flow can give indication of instrument returns, at least over short intervals (see also Chordia and Subrahmanyam 2004; Stoll 1978). Experienced and vigilant market participants (perhaps trading with algorithms) are likely to notice this level of deviation from random walk benchmarks. They are likely to tender market orders with the aim of profiting from the arbitrage. The choice of market orders is informed by the need to quickly profit before the arbitrage opportunity disappears, as this would likely be fleeting. The submitted orders from the arbitrageurs, assuming they are made in ample volumes and on time, are the ones that would lead to relieving the pressure on the market makers inventories. This then potentially spurs the correction of the asset prices. According to Chordia et al. (2005), the correction in asset prices decreases return predictability. Since arbitrage traders are more likely to tender these orders when the spreads are narrow (e.g. see Brennan and Subrahmanyam 1998 for the influence of liquidity on trading tactics; Peterson and Sirri 2002), one would expect reduced return predictability when the market is fairly liquid than otherwise. The interrelatedness of liquidity, price discovery, market efficiency and trade impact is therefore evidenced.

Preceding paragraphs discuss findings based on analyses conducted on equity and more traditional financial market platforms, since those instruments/platforms have been extensively investigated. However, transaction costs-related contributions on exchange-traded emissions permits using high-frequency data have been scarce. Many general contributions to the transaction cost literature in environmental finance and economics have been to upstream issues such as initial allocation of emission permits and market conception. Convery (2009) provides a broad overview of a number of the early studies (see also Burtraw et al. 2011). The body of literature in carbon trading–related finance is growing although very few are focused on market microstructure or pricing at the intraday level. A number of studies have been aimed at understanding general financial market characteristics of the EU-ETS. A significant proportion of these studies is based on trading in the first (trial) phase of the EU-ETS, the Phase I. The following paragraphs briefly discuss some of the related studies.

Benz and Hengelbrock (2009) are the first to provide an intraday analysis of liquidity and price discovery in the European carbon futures market. They investigate transaction costs in the now-agreed largely inefficient Phase I of the EU-ETS. They use the Engle and Granger (1987) vector error correction model (VECM) framework to determine the price process leader between two major platforms in the EU-ETS; they show that the European Climate Exchange (ECX) leads the price discovery process ahead of Nord Pool. Their paper is important in that they employ intraday data. Mizrach and Otsubo (2014) also investigate the initiation of price discovery between two platforms, this time between the Bluenext spot market in Paris and the ECX futures market in London. They use Hasbrouck (1995) and Gonzalo and Granger (1995) information share estimation approaches. They report that the ECX is responsible for about 90% of the share of combined price discovery for the two platforms.

In a study using identical methodologies to Benz and Hengelbrock (2009), Rittler (2012) examines price discovery and causality issues in the early part of Phase II. The study aims to identify the price leader between the spot traded from Bluenext, Paris, and futures contracts traded on the ECX, London. The two studies are similar not only in techniques but also in the direction of investigation. Benz and Hengelbrock (2009) focus on price leadership between two platforms (ECX and Nord Pool) in the EU-ETS, and Rittler (2012) on price leadership between two instruments (spot and futures contracts). Cason and Gangadharan (2011) conduct a laboratory examination of price discovery in linked emissions trading markets. They find improvements in price discovery and efficiency as a result of intermediation between linked markets. This holds some relevance to this book, because the EU-ETS has already been linked to countries outside the EU (Iceland, Liechtenstein and Norway), and European Union Allowances (EUAs) created by those countries are traded on the ECX platform. Finally on liquidity, Frino et al. (2010) examine liquidity and transaction costs in the EU-ETS using intraday data from the ECX using quarterly computations; their results show general improvement in liquidity over the course of Phase I and during the first two quarters of Phase II.

There are several studies relating to Rittler (2012) in his examination of links between the spot and futures contracts. Uhrig-Homburg and Wagner (2007) employ daily data from Phase I to determine price discovery measures for both spot and futures; the study suggests that futures lead the price discovery process. Daskalakis et al. (2009) explore the links between spot and futures by modelling EUA price dynamics using stochastic processes in Phase I. They find that interphase banking restrictions are associated with inconsistencies in futures pricing during Phase I. Futures pricing only conforms to the cost of carry model on intraphase basis. This finding is supported to some extent by Joyeux and Milunovich (2010) since they show that long-run links exist between spot and futures in Phase I, they report intertemporal links for spot with two futures contracts tested.

Daskalakis and Markellos (2008) earlier show that the market in Phase I did not conform to weak-form efficiency. The authors suggest that the lack of efficiency could be due to banking restrictions and immaturity of the market in Phase I. Montagnoli and de Vries (2010) concur with this assessment of the Phase I market. They show that Phase I was an inefficient experiment, with thin trading leading to a huge bias for the efficient market hypothesis. The study goes on to examine trading in the very early period of the second phase and report significant improvements in market efficiency.

A different stream of literature studies several factors as contributors to price formation: Christiansen and Arvanitakis (2005), Mansanet-Bataller et al. (2007), Alberola et al. (2008) and Bredin and Muckley (2011) using daily data, explore the effect of changes in energy fundamentals on daily EUA returns in the EU-ETS. Their reports suggest that pricing in the EU-ETS is driven by the fundamentals tested. Miclăuş et al. (2008) apply event study using the AR(1)-GARCH(1,1) model to examine the effects of regulatory events such as National Allocation Plan (NAP) and Voluntary Emission Reduction (VER) announcements on EUA prices. They report that not all cumulated abnormal returns obtained are statistically significant; hence, market participants in Phase I were able to anticipate future market conditions. Mansanet-Bataller and Pardo Tornero (2007) also apply an event study methodology to examine the effects of regulatory events; their results imply that regulatory and policy announcements have an impact on EUA prices in Phase I. Indeed, Hintermann (2010), using a market that expresses permit prices as a function of several variables, suggest that prior to a price crash that occurred in April 2006,[1] the price of carbon permits was driven by policy rather than marginal abatement costs. Fezzi and Bunn (2009) employ a VAR procedure and present results implying that electricity prices are jointly influenced by shocks in carbon prices. Their findings suggest that a 1% increase in the price of emission permits resulted in an increase of 0.32% in UK electricity prices during Phase I. Nazifi and Milunovich (2010) also report links between electricity and emission permits' prices (as well as between emission permits and natural gas, and emission permits and oil); however, their results also indicate that there is no evidence of a long-run relationship between carbon prices and coal, oil, natural gas and electricity.

This implies that the prices are not cointegrated and may wander apart without bounds in the long run.

The foregoing discussion sets the background for this book's examination of the interrelated issues of price discovery, liquidity, pricing efficiency and policy impacts in the EU-ETS. Since the preponderance of EU-ETS studies (as discussed in the last four paragraphs) is based on an examination of trading in Phase I, and the third phase is still a work in progress, this book is focused on Phase II of the EU-ETS. There is indeed a huge literature gap on intraday evolution of price discovery and liquidity in both Phases I and II. If the markets have tended towards maturity in Phase II, one would expect minimal serial dependence over the day; hence in order to infer on microstructure properties of the market, intraday analysis of the market microstructure is needed. There are very few studies addressing this gap at the moment. This book attempts to fill the gap by presenting new results and also by bringing together the relevant existing studies.

Note

1. The price collapse appears to have been triggered by an unorganised release of the first set of emission verification results in Phase I. The incident is discussed in Hintermann (2010), Daskalakis et al. (2011) and Bredin et al. (2011).

References

Acharya, V. V., & Pedersen, L. H. (2005). Asset Pricing with Liquidity Risk. *Journal of Financial Economics, 77*, 375–410.

Alberola, E., Chevallier, J., & Chèze, B. (2008). Price Drivers and Structural Breaks in European Carbon Prices 2005–2007. *Energy Policy, 36*, 787–797.

Amihud, Y. (2002). Illiquidity and Stock Returns: Cross-section and Time-series Effects. *Journal of Financial Markets, 5*, 31–56.

Amihud, Y., Mendelson, H., & Lauterbach, B. (1997). Market Microstructure and Securities Values: Evidence from the Tel Aviv Stock Exchange. *Journal of Financial Economics, 45*, 365–390.

Benz, E., & Hengelbrock, J. (2009). *Price Discovery and Liquidity in the European CO_2 Futures Market: An Intraday Analysis*. Paper Presented at the Carbon Markets Workshop, 5 May 2009.

Bredin, D., Hyde, S., & Muckley, C. (2011). *A Microstructure Analysis of the Carbon Finance Market*. University College Dublin Working Paper, Dublin.

Bredin, D., & Muckley, C. (2011). An Emerging Equilibrium in the EU Emissions Trading Scheme. *Energy Economics, 33*, 353–362

Brennan, M. J., & Subrahmanyam, A. (1996). Market Microstructure and Asset Pricing: On the Compensation for Illiquidity in Stock Returns. *Journal of Financial Economics, 41*, 441–464.

Brennan, M. J., & Subrahmanyam, A. (1998). The Determinants of Average Trade Size. *The Journal of Business, 71*, 1–25.

Burtraw, D., Goeree, J., Holt, C., Myers, E., Palmer, K., & Shobe, W. (2011). Price Discovery in Emissions Permit Auctions. In R. M Isaac & D. A. Norton (Eds.), *Research in Experimental Economics* (pp. 11–36). Bingley, UK: Emerald Group Publishing..

Cason, T. N., & Gangadharan, L. (2011). Price Discovery and Intermediation in Linked Emissions Trading Markets: A Laboratory Study. *Ecological Economics, 70*, 1424–1433.

Chan, L. K. C., & Lakonishok, J. (1993). Institutional Trades and Intraday Stock Price Behavior. *Journal of Financial Economics, 33*, 173–199.

Chang, Y. Y., Faff, R., & Hwang, C.-Y. (2010). Liquidity and Stock Returns in Japan: New Evidence. *Pacific-Basin Finance Journal, 18*, 90–115.

Chiyachantana, C. N., Jain, P. K., Jiang, C., & Wood, R. A. (2004). International Evidence on Institutional Trading Behavior and Price Impact. *The Journal of Finance, 59*, 869–898.

Chordia, T., Roll, R., & Subrahmanyam, A. (2005). Evidence on the Speed of Convergence to Market Efficiency. *Journal of Financial Economics, 76*, 271–292.

Chordia, T., Roll, R., & Subrahmanyam, A. (2008). Liquidity and Market Efficiency. *Journal of Financial Economics, 87*, 249–268.

Chordia, T., & Subrahmanyam, A. (2004). Order Imbalance and Individual Stock Returns: Theory and Evidence. *Journal of Financial Economics, 72*, 485–518.

Christiansen, A. C., & Arvanitakis, A. (2005). Price Determinants in the EU Emissions Trading Scheme. *Climate Policy, 5*, 15–30.

Chung, D. Y., & Hrazdil, K. (2010). Liquidity and Market Efficiency: A Large Sample Study. *Journal of Banking & Finance, 34*, 2346–2357.

Convery, F. J. (2009). Reflections—The Emerging Literature on Emissions Trading in Europe. *Review of Environmental Economics and Policy, 3*, 121–137.

Cushing, D., & Madhavan, A. (2000). Stock Returns and Trading at the Close. *Journal of Financial Markets, 3*, 45–67.

Daskalakis, G., Ibikunle, G., & Diaz-Rainey, I. (2011). The CO_2 Trading Market in Europe: A Financial Perspective. In A. Dorsman, W. Westerman, M. B. Karan, & Ö. Arslan (Eds.), *Financial Aspects in Energy: A European Perspective* (pp. 51–67). Berlin; Heidelberg: Springer.

Daskalakis, G., & Markellos, R. N. (2008). Are the European Carbon Markets Efficient? *Review of Futures Markets, 17,* 103–128.

Daskalakis, G., Psychoyios, D., & Markellos, R. N. (2009). Modeling CO_2 Emission Allowance Prices and Derivatives: Evidence from the European Trading Scheme. *Journal of Banking & Finance, 33,* 1230–1241.

Datar, V. T., Naik, N. Y., & Radcliffe, R. (1998). Liquidity and Stock Returns: An Alternative Test. *Journal of Financial Markets, 1,* 203–219.

Easley, D., & O'Hara, M. (1987). Price, Trade Size, and Information in Securities Markets. *Journal of Financial Economics, 19,* 69–90.

Engle, R. F., & Granger, C. W. J. (1987). Co-integration and Error Correction: Representation, Estimation, and Testing. *Econometrica, 55,* 251–276.

Epps, T. W. (1979). Comovements in Stock Prices in the Very Short Run. *Journal of the American Statistical Association, 74,* 291–298.

Fama, E. F. (1970). Efficient Capital Markets: A Review of Theory and Empirical Work. *The Journal of Finance, 25,* 383–417.

Fezzi, C., & Bunn, D. (2009). Structural Interactions of European Carbon Trading and Energy Prices. *The Journal of Energy Markets, 2,* 53–69.

Foster, F. D., & Viswanathan, S. (1993). Variations in Trading Volume, Return Volatility, and Trading Costs: Evidence on Recent Price Formation Models. *The Journal of Finance, 48,* 187–211.

Frino, A., Kruk, J., & Lepone, A. (2010). Liquidity and Transaction Costs in the European Carbon Futures Market. *Journal of Derivatives and Hedge Funds, 16,* 100–115.

Glosten, L. R., & Harris, L. E. (1988). Estimating the Components of the Bid/Ask Spread. *Journal of Financial Economics, 21,* 123–142.

Glosten, L. R., & Milgrom, P. R. (1985). Bid, Ask and Transaction Prices in a Specialist Market with Heterogeneously Informed Traders. *Journal of Financial Economics, 14,* 71–100.

Gonzalo, J., & Granger, C. W. J. (1995). Estimation of Common Long-memory Components in Cointegrated Systems. *Journal of Business and Economic Statistics, 13,* 27–35.

Grossman, S. J., & Miller, M. H. (1988). Liquidity and Market Structure. *The Journal of Finance, 43,* 617–633.

Hasbrouck, J. (1991a). Measuring the Information Content of Stock Trades. *The Journal of Finance, 46,* 179–207.

Hasbrouck, J. (1991b). The Summary Informativeness of Stock Trades: An Econometric Analysis. *The Review of Financial Studies, 4*, 571–595.

Hasbrouck, J. (1995). One Security, Many Markets: Determining the Contributions to Price Discovery. *The Journal of Finance, 50*, 1175–1199.

Hendershott, T., Jones, C. M., & Menkveld, A. J. (2011). Does Algorithmic Trading Improve Liquidity? *The Journal of Finance, 66*, 1–33.

Hillmer, S. C., & Yu, P. L. (1979). The Market Speed of Adjustment to New Information. *Journal of Financial Economics, 7*, 321–345.

Hintermann, B. (2010). Allowance Price Drivers in the First Phase of the EU ETS. *Journal of Environmental Economics and Management, 59*, 43–56.

Holthausen, R. W., Leftwich, R. W., & Mayers, D. (1990). Large-Block Transactions, the Speed of Response, and Temporary and Permanent Stock-price Effects. *Journal of Financial Economics, 26*, 71–95.

Joyeux, R., & Milunovich, G. (2010). Testing Market Efficiency in the EU Carbon Futures Market. *Applied Financial Economics, 20*, 803–809.

Kraus, A., & Stoll, H. R. (1972). Price Impacts of Block Trading on the New York Stock Exchange. *The Journal of Finance, 27*, 569–588.

Kyle, A. S. (1985). Continuous Auctions and Insider Trading. *Econometrica, 53*, 1315–1335.

Lee, C. M., & Ready, M. J. (1991). Inferring Trade Direction from Intraday Data. *The Journal of Finance, 46*, 733–746.

Mansanet-Bataller, M., Pardo, T., & Valor, E. (2007). CO_2 Prices, Energy and Weather. *The Energy Journal, 28*, 73–92.

Mansanet-Bataller, M., & Pardo Tornero, Á. (2007). *The Effects of National Allocation Plans on Carbon Markets*. University of Valencia Working Paper, Valencia.

Miclăuş, P. G., Lupu, R., Dumitrescu, S. A., & Bobircă, A. (2008). Testing the Efficiency of the European Carbon Futures Market using the Event-Study Methodology. *International Journal of Energy and Environment, 2*, 121–128.

Mizrach, B., & Otsubo, Y. (2014). The Market Microstructure of the European Climate Exchange. *Journal of Banking & Finance, 39*, 107–116.

Montagnoli, A., & de Vries, F. P. (2010). Carbon Trading Thickness and Market Efficiency. *Energy Economics, 32*, 1331–1336.

Nazifi, F., & Milunovich, G. (2010). Measuring the Impact of Carbon Allowance Trading on Energy Prices. *Energy & Environment, 21*, 367–383.

O'Hara, M. (2003). Presidential Address: Liquidity and Price Discovery. *The Journal of Finance, 58*, 1335–1354.

Pástor, L., & Stambaugh, R. F. (2003). Liquidity Risk and Expected Stock Returns. *The Journal of Political Economy, 111*, 642–685.

Patell, J. M., & Wolfson, M. A. (1984). The Intraday Speed of Adjustment of Stock Prices to Earnings and Dividend Announcements. *Journal of Financial Economics, 13*, 223–252.

Peterson, M., & Sirri, E. (2002). Order Submission Strategy and the Curious Case of Marketable Limit Orders. *The Journal of Financial and Quantitative Analysis, 37*, 221–241.

Rittler, D. (2012). Price Discovery and Volatility Spillovers in the European Union Emissions Trading Scheme: A High-frequency Analysis. *Journal of Banking & Finance, 36*, 774–785.

Stoll, H. R. (1978). The Supply of Dealer Services in Securities Markets. *The Journal of Finance, 33*, 1133–1151.

Uhrig-Homburg, M., & Wagner, M. (2007). Derivative Instruments in the EU Emissions Trading Scheme—An Early Market Perspective. *Energy and Environment, 19*, 1–26.

2

Emissions Trading in Europe: Background and Policy

2.1 Introduction

Emissions trading via cap and trade was established as a viable policy instrument by the work of Montgomery (1972). When employing a static framework under perfect market conditions, one may observe that, for participating firms in an emissions-constrained economy, there is a minimum cost equilibrium. Rubin (1996), using optimal control theory, also shows that the equilibrium price for greenhouse gas (GHG) emission permits corresponds to the costs due to marginal abatement if firms are allowed to bank and borrow emission permits (see Springer 2003 for a review of results gathered from 25 models on marketable emission permits). Emissions trading thus provides the avenue for firms to fund emissions reduction programmes through the sale of excess permits earned by investing in abatement measures in the first place. Also, firms are able to reduce emissions by applying the most economical means of doing so. Emissions trading as a component of first, the Kyoto Protocol, and then the subsequent climate treaties, continues to play an important role in the drive to reduce global GHG emissions. The Kyoto Protocol, an international framework with 192 parties, came into force in January 2005. The framework spurred the growth of a multi-billion-dollar sector for emissions trading. The European carbon market has accounted for driving more than 95% of global market share by value for any given year since 2006.

The Kyoto Protocol provides three mechanisms with which participating countries can achieve their emissions targets. The mechanisms include the International Emissions Trading (IET), the Clean Development Mechanism (CDM) and Joint Implementation (JI). The three mechanisms provided for by the Kyoto Protocol are representative of the different statuses of its signatories. Under Kyoto Protocol, 37 industrialised countries and the EU accepted binding reduction targets over a five-year period (2008–2012). The parties can use the mechanisms provided in achieving their individual targets. Article 4 of the Protocol allows for parties to accept emission reduction targets as part of a bubble. The EU, in accepting emission reduction targets as a unit, is considered a bubble. Trading currently takes place electronically with the setting up of an International Transaction Log (ITL), which has been operational since 2007. The ITL remains operational today under the mandate of the Doha amendment to the Kyoto Protocol and other subsequent climate treaties.

IET between countries requires that both nations must have accepted Kyoto Protocol sanctioned caps or abide by the conditions set under subsequent climate treaties. For JI, however, both must have signed on the Protocol in principle but need not have accepted caps in GHGs emissions. Nations with economies in transition to a market economy can be helped to attain energy efficiency through the JI mechanism. The JI is essentially a mechanism for helping nations with economies in transition to a market economy attain a competitive level of energy efficiency. JI thus aids the flow of investment in green technology and projects from Western Europe to the eastern part of the continent.

Although some developing countries are bigger emitters of GHGs than developed countries, those countries did not accept caps under the Kyoto Protocol. This was a particularly controversial aspect of the Kyoto Protocol, especially in relation to China which was and remains the largest emitter of GHGs but is still regarded as a 'developing' country. The absence of emissions reduction quotas for developing nations under the Kyoto Protocol is due to the fact that, historically, industrialised countries are responsible for most of the current global GHG stock. The initial lack of quotas for developing countries is therefore in keeping with the fairness doctrine of the United Nations Framework Convention on Climate Change (UNFCCC). By virtue of their status under the Kyoto Protocol, developing nations are not eligible for IET and JI; hence, the design of CDM to act as a channel for transfer of green funds and clean technologies to developing countries. However, it has been argued that this effectively encourages unchecked migration of emissions from capped countries to uncapped ones such as the developing countries. Countries that rejected caps, despite being industrialised, such as the United States, may also benefit from this loophole, especially if they offer competitive investment destinations on account of having no emissions constraints.

Under the Kyoto Protocol, the EU aimed to reduce its emissions by 8% below 1990 levels by employing mainly the IET mechanism. Since the lapse of the protocol, new targets have been set and announced. These are subsequently discussed in this book. Beyond the Kyoto Protocol, Europe continues to dominate the emissions permit market and will do so for the foreseeable future; this book therefore focuses on EU-ETS platforms. It is believed that the future of global cap and trade policy is dependent on the success or otherwise of the EU-ETS. The EU-ETS is designed as a compulsory cap and trade scheme where participating installations have a legal requirement to lower their emissions in accordance to set caps. However, these participating installations have the opportunity to buy emissions permits to offset exceeding those caps. The tri-status nature of the Kyoto Protocol has already been embedded in the EU-ETS by the use of the EU Linking Directive (Directive 2004/101/EC). This directive creates a direct 'link' between EU emissions and global climate policy, essentially ensuring that emission credits from the CDM and JI can be submitted in lieu of emission reductions towards the EU's reduction target.

The EU-ETS is both the largest compulsory cap and trade scheme in the world and the most potent regional climate change policy tool arising from the EU's 2002 ratification of the Kyoto Protocol, a global treaty on GHG emission reduction. The operation and success of the EU-ETS will significantly inform the direction of global climate policy, identifying effective mechanisms for carbon trading and the effect of different restrictions on the price of emissions. The scheme may also affect the growth of the emission-constrained economies in Europe, New Zealand, the United States, Japan and other parts of the world. This is because regulatory arbitrage is likely to be greater when carbon trading is limited to significantly smaller geographical locations, as is the case at this time. The market is artificial and dependent on environmental policy and regulation; it is therefore exposed to greater levels of uncertainty than is the case for most 'natural' commodities. Emissions permits have to be surrendered on an annual basis. In principle, therefore, futures contracts on emissions permits offer significant benefits, both as instruments for hedging price risk and as mechanisms to assist in the smooth operation of the system as a whole.

The use of permit trading is not novel to the Kyoto Protocol or the EU-ETS. Until the start of the EU-ETS, the most prominent example of emissions permit trading has been the United States Acid Rain programme. The Environmental Protection Agency (EPA) has employed emissions trading as a policy tool to achieve emissions reductions since 1992. There is no shortage of publications examining the operational value of this programme with most concurring on its success. Joskow et al. (1998) and Albrecht et al. (2004) concur on the efficiency of the SO_2 market and the fact that it is a market that meets its target of reigning in SO_2 emissions. The allowances are traded as spot, forward and options, underscoring its liquidity.

A number of theoretical and empirical contributions provide evidence of the EU-ETS's market characteristics; hence, the nature of this market is becoming quite apparent. However, some EU-ETS phase-dependent issues remain largely unexplored. This chapter provides a further discussion of these issues from economic, policy and financial regulation perspectives as a background to the empirical investigations presented in this book. The issues reviewed span the first two phases of the scheme (2005–2012). In subsequent chapters, issues related to the current (third) phase are examined.

2.2 Trading Emissions Under Cap and Trade

The design of cap and trade is hinged on regulating aggregate emissions by putting a price on a target gas. The permission to emit the aforementioned regulated gas is issued as allowances. Emissions trading contrasts taxation in one respect, the regulator decides only a cap, and then the market sets the price on the target gas based on scarcity induced by the regulator's cap (supply) and aggregate demand for the right to emit the gas. In such a scheme, the regulator is not directly responsible for fixing the price of emissions; however, they can influence the pricing of the targeted gas through the introduction of policies aimed at controlling supply, as well as demand. For example, in the EU-ETS, there is a percentage of emission allowances held in reserve (called new entrants reserve) that can be allocated in order to influence pricing. The recently introduced Market Stability Reserve (MSR), aimed at shoring up falling prices during the third phase of the scheme, is also another example of such policies. Permission to emit the regulated gas is issued as allowances. The issuance of the allowances can take the form of free allocation based on historical emissions patterns of an industry (grandfathering), through auction or a combination of both to any degree. The companies and installations involved in this scheme reserve the right to trade their emission allowances, giving them market-based options on meeting the volume targets set by the regulator. A company can be a net seller if it finds it to be more cost-effective to reduce emissions than to buy allowances; it can then sell off its surplus allocations to participants with excess emissions. On the other hand, if purchasing allowances improve firm competitiveness, the firm may opt to be a net buyer (see Boemare and Quirion 2002; Böhringer and Lange 2005).

2.3 Structure of the EU-ETS

For the time being, there are four identified phases of the EU-ETS. The three-year pre-Kyoto commitment trial period (2005–2007) is the so-called Phase I, while the Kyoto commitment period itself (2008–2012) is known as Phase II. The EU decided to proceed to a third phase starting from 2013 to 2020, even while there was no global agreement on a successor to the Kyoto Protocol at the time. The climate change treaty signed by the conference of parties in Durban, South Africa in December 2011 affirmed that decision. Phase IV is expected to start in 2021 and end in 2030.

The EU-ETS is the main element of the EU's climate change policy. The EU adopted a 'burden sharing agreement' (Council Decision 2002/358/CE) allowing it to reallocate emissions reduction targets within its member states so as to avoid stifling vital economic growth in the less wealthy countries. This consequently leads to the setting of more demanding targets for the larger European economies such as Germany and the United Kingdom. The member countries are responsible for identifying the installations affected within their borders under the aggregated reduction target. Iceland, Liechtenstein and Norway (non-EU European countries) also participate in the EU-ETS.

About 12,000 installations with a minimum generating capacity of 20 megawatts (MW) in a number of sectors within the EU operate under EU-ETS rules.[1] Combined, these installations account for approximately 40% of the EU's total GHG emissions (see Hawksworth and Swinney 2009), with 2.2 billion tCO_2 distributed to the affected installations in Phase I alone (see Dodwell 2005). There are several key differences in the operational features of the phases of the EU-ETS, and these are summarised in Table 2.1.

2.3.1 Emissions Permits Creation and Use

For each phase of the scheme, each participating country develops a National Action Plan (NAP) stating the reduction in GHGs emission required for each installation over a stated period as per the burden sharing agreement. The amount of emission permits *created* is subject to approved emissions quota in the plans. The NAP thus also creates a finite amount of emission permits called *European Union Allowances (EUAs)*. Each EUA grants the holder the right to emit one tonne of CO_2 (tCO_2). In the first two phases of the EU-ETS, only CO_2 emissions are targeted. By creating a limited volume of EUAs, a cap is placed on the volume of CO_2 emissions during the period.

At firm level, in Phase I, 95% of EUAs created was allocated free to installations and 5% held in reserve for new entrants (the new entrant's reserve is aimed at providing some period of flexibility for new companies and installations entering into the scheme). In Phase II, up to 10% of the total available was auctioned across EU countries to various degrees. In the

Table 2.1 Phases of the European Union Emissions Trading Scheme (EU ETS)

Variables	Phase I	Phase II	Phase III	Phase IV
Gases targeted	CO_2 only	CO_2 only	CO_2, perfluorocarbons and nitrous oxide	
Allocation system	Allocation based on grandfathering	Up to 10% of created permits can be auctioned, the balance auctioned or held in reserve for new entrants	20% auctioning in 2013 with gradual rise to 70% by 2020. The degree of auctioning varies by industry and country	Linear reduction factor for auctioned percentage is set at 2.2% from 2021 onwards
Proportion of greenhouse gases under scheme	40%	40%	50%	50%
Banking regulations	Intraphase banking only (France and Poland allowed conditional banking to Phase II)	Interphase banking allowed	Interphase banking allowed	Interphase banking allowed
Allocation planning	National allocation plans	National allocation plans	European Union-wide allocations	European Union-wide allocations
Penalty for default	€40 per EUA not submitted plus submission of missing EUA in subsequent compliance year	€100 per EUA not submitted plus submission of missing EUA in subsequent compliance year	Penalty per EUA not submitted is aligned with the European price index, plus submission of missing EUA	Penalty per EUA not submitted is aligned with the European price index, plus submission of missing EUA

The table compares the three phases of the EU-ETS by using regulatory issues as the basis of comparison. Phase I ran between 2005 and 2007, Phase II ran between 2008 and 2012, Phase III started in 2013 and will run until 2020, while Phase IV is planned for a 2021 start and is expected to end in 2030.

third phase (Phase III), EU member states auction an increasing proportion of EUAs, starting from 20% of the total stock of EUAs in 2013 and increasing it annually in order to reach 70% by 2020. The eventual objective is to attain full auctioning at the firm level by 2027 post-Phase III (see Directive 2009/29/EC).

All EU national registries are linked and are connected to the European Commission's Community Independent Transaction Log (CITL), which chronicles all changes to EUA rights by stakeholders at both national and continental levels. Consequently, the national registries record of firm-level trading activities is replicated on the CITL as every trade within a member state is registered at a relevant national registry. Annually, by March 31, all installations under the EU-ETS are required to provide an externally audited report to the relevant emissions regulator. This report will include the volume of CO_2 emitted in the course of operations during the preceding compliance year. By April 30, the installations then have to submit an amount of EUAs equal to the volume emitted during the year. The submission actually involves the deletion of the appropriate number of EUAs from the installations' accounts with the relevant registries.

There are measures dealing with failure to comply: In Phases I and II, the penalty includes €40 and €100 fine, respectively, for each EUA not surrendered for deletion as well as the eventual surrendering of the EUA during the next compliance year. In Phase III, the penalty will see increases based on the European index of consumer prices (see Directive 2009/29/EC). This new provision amends Article 16:4 of the EU Directive 2003/87/EC. As with the two previous phases, the missing EUAs will still be delivered in the next compliance year, the penalty notwithstanding. The provision that requires submission of missing EUAs, along with the respective penalties, serves to ensure that a price ceiling is not inadvertently set for EUAs. This ensures that the markets still get to determine EUA prices without the regulator's interference as the weight of penalties may be construed as a signal of the regulator's projected upper bounds for the market and hence may limit potential price appreciation. The European Commission has repeatedly maintained that EUA prices will be determined purely by demand and supply forces.

Project emissions permits are also allowed for submission in the EU-ETS. The 'linking directive' (Directive 2004/101/EC) sanctions that Certified Emission Reduction units (CERs) and Emission Reduction Units

(ERUs) from CDM and JI project sources respectively can be surrendered in place of EUAs in a ratio of 1:1 for compliance purposes. However, with very strict percentage-based limits.

Based on the limits set in the respective member states' NAPs, EUAs are annually grandfathered (grandfathered and auctioned in Phases II, III) to affected installations. This annual allocation usually takes place by the end of February, giving a two-month interval between allocation and the time when installations provide their annual emissions reports. Ideally, the allocations are for the future compliance year; the implication, however, is that the installations can borrow future permits to offset the preceding year's compliance (see Daskalakis et al. 2011). Borrowing future EUAs is permitted for a year only and must be within the phase. EUAs can be banked (stored for future use) from previous years for future compliance. This regulation applies only to Phases II and III of the EU-ETS on an EU-wide scale.

During Phase I, most member states prevented the banking of allowances from 2007 to 2008. This is logical owing to the fact that emissions reductions during 2005–2007 (Phase I) could not be used in achieving the Kyoto commitment period targets. The EU member states were at liberty to allow banking between Phases I and II of the EU-ETS. However, only France and Poland allowed this with some strong restrictions. Nevertheless, the decision to allow banking in every case was still subject to the European Commission's authorisation and banking limits (see MEMO/06/452 of the European Commission). Specifically, in Poland and France, installations were allowed to bank the maximum of the difference between their primary allocation and their verified emissions. The permits acquired on the market were not eligible for banking.

2.4 Critical Phase-Dependent Issues Arising from EU-ETS Design and Regulations

2.4.1 The Problem with Ban on Intertemporal Trading

The ban placed on the transfer of permits from Phase I to Phase II mentioned in Sect. 2.3.1 and highlighted in Table 2.1 does have some far-reaching implications; the discussions here lead up to these consequences.

A number of studies, such as Joskow et al. (1998), Schmalensee et al. (1998), Stavins (1998), Butzengeiger et al. (2001) and Svendsen and Vesterdal (2003), examine the design and market structure of cap and trade from economics and policy viewpoints. From these studies, it is clear that the success of the cap and trade system is predicated on a number of factors. Liquidity, diversity of industries involved and flexibility for firms to develop emission reduction strategies are vital requirements for a successful scheme. Firms must be able to reduce emissions only when it is more economically efficient to do so rather than purchase emission permits in the market. The overriding consideration should be to ensure compliance with capped levels at the least possible cost, hence the need for flexibility (see also Rubin 1996; Kling and Rubin 1997; Schennach 2000; Schleich et al. 2006). A greater level of efficiency and success of the system can be ensured when the scheme permits the transfer of emissions allowances between periods (intertemporal trading). This means allowing firms to borrow permits from future periods and transfer excess permits to future periods for the purpose of compliance. Despite the near-unanimity of this perspective among academics prior to the start of the EU-ETS, the EU opted to ban intertemporal trading between Phase I and Phase II of the EU-ETS. A major reason was that the first phase of the scheme was just a trial phase aimed at fine-tuning the scheme for the Kyoto commitment phase (Phase II). Had banking of allowances been permitted, the EU's Kyoto target may not have been achieved with excess liquidity building in Phase I. However, allocations in Phase II could have been made to account for the excess quantity available from Phase I, helping avert excess liquidity and market quality decline issues in both phases. The decision not to allow intertemporal trading eventually led to significant losses in market value for derivative instruments with delivery dates near the close of Phase I. This is because they have no market value in the next phase (see Daskalakis et al. 2009).

2.4.2 The Problem with Initial Allocation in Phase I

The method of allocation of start-up allowances on ETSs has always been a hotly debated topic since the passing of the US Clean Air act and such debates have only become more contentious with the setting up of the

EU-ETS. In the allocation of initial allowances, the volume released into the system is very crucial in ensuring the credibility of the scheme. Tight caps must be in place to engender relative scarcity, which builds a measure of price responsiveness (see Boemare and Quirion 2002; Vesterdal and Svendsen 2004; Böhringer and Lange 2005; Neuhoff et al. 2006). In Phase I of the EU-ETS, grandfathering was the sole method of allocation, although in theory auctioning is more desirable. Auctioning at a price sends a strong indication of the value placed on the allowances by the regulator and fosters price signal transparency. Price transparency improves information dissemination and builds confidence of a clear market structure and a transparent price discovery process (see Grubb and Neuhoff 2006; Neuhoff et al. 2006). Auctioning is also a revenue generator. Revenue generation can give governments more options in discretionary spending. For example, revenues from auctions could fund research and development in green technology and, in the event of emissions leakage, vulnerable industries could also be supported to retain competitiveness.

The release of CO_2 into the atmosphere is a restricted activity in the EU, thus the EU economy is 'emissions-constrained'. This constraint translates into costs for firms through the purchase of permits. The cost of permits in a pure economic view should be passed on to consumers. The power sector in the EU is the sector most impacted by the EU-ETS. If the costs of permits are passed on to consumers, it is expected that power prices will increase in relation to the costs of those permits (see Linares et al. 2006; Kara et al. 2008; Reneses and Centeno 2008). However, since EUA levels were net long due to over-allocation of allowances in Phase I, no increases in costs should occur between 2005 and 2007 on account of the EU-ETS. In addition, allocations were made to the companies involved through grandfathering, firms that incurred no costs during the allocation process. It is understood that costs could still be incurred under grandfathering, but only if EUAs had been net short, which they were not. Ceteris paribus Phase I should not have caused price increases in the power market. However, a number of studies report price increases in electricity costs to the consumer, supposedly as a consequence of carbon costs. For example, Fezzi and Bunn (2009) report that in Phase I, a 1% increase in the price of carbon resulted in an increase of 0.32% in UK electricity prices (see also Sijm et al. 2005, 2006). Thus, it appears that the free allocation of emission permits led to a financial windfall for electricity producers.

The general conclusion that can be made on the allocation mechanism in Phase I is simply that it fell short of desired results. The policy aim of reducing emissions through changes in institutional and individual power use contingent on rising costs of power was largely not achieved. The impact of carbon prices on electricity prices, however, suggests that this is achievable. The allocation mechanism in Phase I was ill-advised and poorly executed, and this led to installations having no use to invest in emissions abatement measures. Having stated this, Phase I must have had some impact on emissions in the latter stages because the emission reports for 2007 according to the European Commission show a 1.6% decline in EU aggregate emissions. The approaching economic slowdown of 2008 could have also contributed to the decline in emissions. Auctioning or a combination of both auctioning and grandfathering are the preferred alternatives to pure grandfathering (see Boemare and Quirion 2002; Böhringer and Lange 2005; Sijm et al. 2006). A combination approach was adopted for Phase II, with better outcomes than Phase I.

2.4.3 Carbon Price in the EU-ETS

Carbon price captures the incorporation of the costs of emitting carbon into production processes in an emissions-constrained economy (Labatt and White 2007). Theoretically, the price of carbon is expected to be correlated with the marginal costs of abatement of GHGs emissions reduction (see Hintermann 2010; Rubin 1996). Market frictions (from a microstructure perspective) and exogenous impacts such as restrictions on banking and borrowing (from a policy perspective) however readily complicate the pricing of carbon; this book examines some of these issues in subsequent chapters.

Analysing emission permits pricing requires an understanding of what EUAs behave like under trading conditions: commodities or other financial asset classes, or perhaps they are different to both classes? As EUAs are electronically generated records on EU member states' registries, they have features that straddle several financial asset classes. Although EUAs share features with commodities and they can be regarded as commodities in terms of delivery as an underlying of derivative contracts, delivery of EUAs is virtually 'risk-free' in comparison with more traditional commodities such as agricultural produces and petroleum products (Daskalakis et al. 2011). The transfer of EUAs only requires a recorded transfer from one holder to the other at the national registry, hence are traded and delivered in a way similar to regular financial futures. Under close scrutiny, EUAs exhibit characteristics of traditional commodities in that classic supply and demand influences are fundamental drivers of their price structure. As with many commodities, EUAs can also be classed as a factor in production. Historically, demand for EUAs is driven by the demand for energy (see Alberola et al. 2008) and factors that also influence the demand for other, more traditional, commodities. These include factors such as the state of the economy, lifestyle and changes in consumers' tastes, and so on. The supply end of the EUA is almost entirely influenced by policies of regulating agencies. International treaties based on scientific developments then in turn influence regulators.

With this rather slightly more complex web of non-market factors, carbon pricing should hardly correlate with financial asset classes. Daskalakis et al. (2009) explore the question of correlation of CO_2 emission allowances with other asset classes and find no significant correlation with major asset classes; their results therefore suggest that carbon financial instruments could be used for investment diversification. There are striking similarities between CO_2 emission allowances and financial assets that cannot be overlooked, however. For example, as it is with financial assets, the possession of CO_2 emission allowances endows a right to an underlying benefit. In the case of EUAs, the benefit is the permission to pollute to a certain degree. Also, similar to financial assets, EUAs are not liable to physical losses, they need not be physically stored, they have unrestricted tradability and their pricing can be determined by supply and demand. Regulations can influence upper bounds of EUAs, and this can force the development of different properties from those of traditional financial instruments. The fundamental characteristic affected by regulations is the pricing of carbon, thus exhibiting stochastic properties at variance with other financial instruments/assets. In contrast to financial asset classes such as bonds and stocks, interests or dividends cannot be paid on CO_2 certificates.[2] Primarily, CO_2 emission allowances are instruments of compliance; as a result, their pricing should reflect the cost of compliance (see Hintermann 2010). Although energy-related variables represent the strongest link in the determination of compliance costs, there are suggestions that linkages may exist between carbon prices and financial markets. For example, Koch (2012) reports a dynamic association between carbon financial instruments and conventional financial assets returns. In Sect. 2.4.4, the empirical evidence on correlations between carbon and financial assets/instruments during the global financial crisis is considered.

2.4.4 Impact of the Global Financial Crisis on the EU-ETS

During the first trading year in Phase II (2008), transactions on the EU-ETS were valued at US $101.49 billion (€74.56 billion) representing an 87% growth rate on the previous year, with more than three billion EUA spot, futures and options contracts traded. In 2008, the recession in Europe and most other parts of the developed world forced significant reduction in the demand for big-ticket items such as housing, cars, and so on. This resulted in a lower demand for cement, steel and other relevant raw materials. Consequently, manufacturing and building projects stalled, leading to lower energy consumption. The need to purchase EUAs is based on energy/power and oil consumption and, as power consumption fell, the demand for EUAs declined. This led to a sharp fall in carbon prices. The spot price during the year plunged 75% over a period of eight months from a record level of €28.73 in July 2008 to a lowly €7.96 on 12 February 2009. The decline was not limited to spot trading; declines as dramatic as the one recorded on spot trades also occurred in the secondary CER market (see Fig. 2.1). Considering this link between carbon permit price and a global economy bellwether commodity, crude oil, it is not surprising that Koch (2012) reports variable correlations between emissions-based futures contracts and financial asset returns.

The decline in demand for CO_2 allowances meant that corporations accumulated more grandfathered allowances than was needed. This, coupled with the tightness in the credit market, led to a huge sell-off by firms

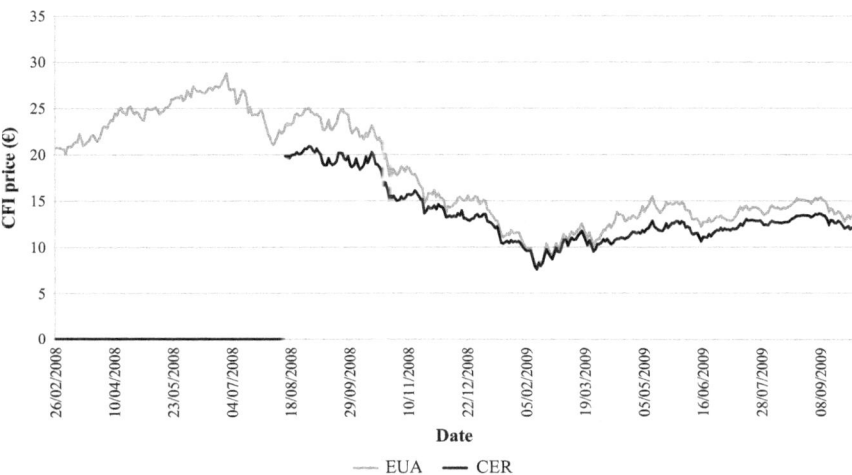

Fig. 2.1 Response of carbon financial instruments to a recession. The figure illustrates the variations in prices of carbon financial instruments trading in the EU-ETS, to the recession of 2008–2009. The chart is plotted with data sourced from BlueNext, Paris. The data spans 26 February 2008 to 30 September 2009

in order to raise funds. The higher rate of supply by these heavy industries, the ditching of long positions by investors (who were not compliance buyers) and the decline in demand in the aftermath of the economic slowdown resulted in a rapid fall in EUA prices and those of its derivatives. During this period, daily and monthly records for spot transactions were broken in a flurry of trading by corporations in search of liquidity in a difficult credit market. Most of the increased trading was reportedly done in the spot market, and this accounted for 36% of all transactions in the EU-ETS in December 2008. This is a major contrast to an initial low of 1% in the first half of the year (Capoor and Ambrosi 2009).

However, as the worst of the recession passed and spare capacity started being used up, consumption began to rise and this led to the gradual return of market confidence. The rising consumer confidence in energy commodities and other asset classes observed in the second half of 2008 was mirrored in emissions permit prices in the EU-ETS, another piece of evidence that carbon prices are correlated with energy commodities (see Christiansen and Arvanitakis 2005; Mansanet-Bataller et al. 2007; Alberola et al. 2008 for reports on carbon price relationship with energy commodities). The price recovery in April 2009 can also be attributed to the need for compliance buyers to submit their EUA allowances for the 2008 compliance period; this generated more trading activity and consequently market liquidity. In 2009, the EU-ETS accounted for 96.46% of global allowances trades (valued at $122.8 billion). This is with a trading volume of around 6.326 billion tCO_2 worth US $118.5 billion (€88.7 billion), up from approximately, 3.1 billion tonnes (worth $101.49 billion) in 2008. The market value in 2009 represents an 18% improvement over 2008; this came despite a 42% dip in EUA prices over the same period. By 2010, the market appeared to have recovered from the recessionary impacts of 2008. In 2010, the value of total EUA traded climbed to US $119.8 billion (more than 84% the global carbon market value), and the EU-ETS driven share of the global market increased to 97% (Linacre et al. 2011).

2.4.5 Regulatory Risk Issues in the EU-ETS

Since 2005, the market has improved remarkably to achieve a level of functionality. However, some issues have come to the fore since the launch of the EU-ETS, which underscore the underlying market risks in the EU-ETS. The first issue relates to the EUA price collapse of April 2006. The over-allocation of permits to ensure that domestic firms retain competitiveness vindicates the concerns already expressed about both market foundation risks and market confidence risks. The price collapse also relates to information asymmetry as a result of disorderly release of information to the market by regulators with little insufficient understanding of the complexities of the market (Frino et al. 2010). Subsequent significant falls in instruments' prices have been recorded in both Phase II and Phase III as well. For example, the EU authorities had to introduce *backloading* and the MSR as measures to stabilise prices and address persistently low EUA prices of around €5/tCO_2.

The second issue is the now infamous Carousel VAT evasion strategy. Many firms took part in this, effectively affording themselves free and temporary funds through culling of VAT imposed on spot trading of EUAs across the EU. The strategy involves the collection of VAT on commodities which is not passed on to the regulatory authorities. In a number of European countries, when buyers and sellers of a commodity are in different jurisdictions, the importers (buyers) are either not subject to paying VAT levies or only required to pay up at a later time, which is usually one to three months. The seller is, however, required to pay the local VAT to the buyer. Conversely, if both the buyer and seller are domiciled in the same country, the VAT is payable by cash on the date of transaction. EUAs exist only as records on national registries and are therefore easily transported across borders. VAT exemptions provide VAT-based funding for the actors, but they also lead to a loss of revenue to the European jurisdictions. In late 2009, Europol estimated that more than €5 billion had been lost by EU countries as a result of the fraudulent practice. Issues such as this greatly affect the integrity of the carbon market/EU-ETS. The most exposed countries, such as France and the United Kingdom, acted to restore market integrity through unilateral measures. These included introducing a reverse charging mechanism and criminal prosecution with arrests made in the United Kingdom, France, Norway and Spain. In September 2009, other measures were proposed by the EU.

The third critical regulatory issue is the recycling of CERs already surrendered for compliance by installations. A significant case is the sale of surrendered CERs by Hungary in March 2010. The EU responded by modifying the registry rules to prevent future occurrences.

Fourth, there have been a number of controversies over submitted NAPs by some countries. The European Commission, exercising its privilege as EU-wide regulator of the EU-ETS, suppressed the issuance of EUAs from the NAPs of Estonia and Poland. These two countries dissented and appealed to the European Court of First Instance and the court promptly annulled the Commission's decision. The authority of the Commission has suffered as a result of this annulment, and this is anticipated to continue to have consequences for EU oversight functions within the scheme.

Finally, one of the biggest risks so far faced by the EU-ETS is insecurity at national registries. In January 2011, it was discovered that EUAs worth more than €45 million had been stolen from certain registries in the EU. This unprecedented discovery led to the temporary closure of some national registries and the suspension of spot trading in the EU-ETS. In the months following the discovery, the European Commission embarked on an EU-wide rehabilitation and improvement of national registry security systems.

The events enumerated above may have dented the credibility of transactions in the EU-ETS, but they have also validated expectations of a maturing market that is becoming fully integrated with the European economy. This is based on the fact that these types of occurrences are common in mainstream conventional markets and have come to be expected in valuable markets, hence the enforcement of financial regulations. Kossoy and Ambrosi (2010, p. 6) argue this point:

> Ironically, however, these controversies provide evidence that the emissions market is maturing and becoming mainstreamed within the European economy. Entities do not seek out loopholes in insignificant markets, fraudsters do not focus on small businesses, and disputes over NAPs demonstrate that carbon has become very important to the involved countries. Through these challenges the EU-ETS has demonstrated resilience and the capacity for swift self-adjustment.

Current international developments underline not only the growing international importance of emissions trading, but also the need for clear evidence on how well the EU-ETS is functioning. Understanding the microstructure of the EU-ETS is therefore important since it can inform policy decisions regarding upcoming IET schemes. Recent events, especially the important deal reached at COP 21 in Paris, suggest that governments in other regions of the world are already progressing towards establishing carbon reduction schemes, with emissions trading remaining the choice of many. In the event that a global ETS is established in order to effectively counter climate change, the scheme is likely to be anchored on the EU-ETS infrastructure if the latter can be proven to be effective. As an indication of developments in the future, three non-EU countries: Liechtenstein, Norway and Iceland have already linked their cap and trade structures to the EU-ETS. In principle, the EU is open to the prospect of interregional linking of cap and trade schemes. Linking various ETS structures around the globe will likely precede a global ETS. The EU-ETS will be vital to this development, especially with respect to provision of required trading activity and liquidity (Hanemann 2009; see e.g., Hobbs et al. 2010).

2.5 Chapter Summary

The EU-ETS is a multi-billion-dollar market deserving of attention from finance and economics researchers. In this chapter, the economic significance of the EU-ETS is examined in detail along with several phase-dependent issues. It also provides a descriptive analysis of the EU-ETS as a background to the empirical studies reported in subsequent chapters of the book. The chapter generally conveys the view that the Phase I experiment in the EU-ETS was not as successful as policy makers had hoped. Phase II proved to be a more successful exercise, although there were challenges. Those challenges have become even more significant in Phase III with persistent low prices across EU-ETS platforms. However, since the EU-ETS remains the main driver for global emissions trading, its success remains critical to any global advances with respect to climate change policy. Studies examined in this chapter suggest that ETSs have the potential to work as informationally efficient financial markets if the right combination of factors is in place, chief among these being the right structure to ensure carbon prices equal to the marginal abatement costs of CO_2 (efficient price signalling). The question then is: Is the EU-ETS (informationally) efficient? Daskalakis and Markellos (2008) provide some answers to this question for Phase I; they suggest that the market did not confirm to weak form efficiency. In the next chapter, I explore the open question of informational efficiency in Phase II using high-frequency data from the largest carbon trading platform in the world with some interesting results. The results here are very important in that they are the first set to show the intraday evolution of the price discovery process for carbon financial instruments. They provide investors, compliance traders and policy makers with critical information required to adequately engage with EU-ETS platforms/the carbon market.

Several other market microstructure issues that can give insights into the state of maturity of the EU-ETS remain largely unexplored for Phase II, and even for Phase I. Liquidity and diversity of industries involved in the EU-ETS are also important in order to maintain market quality (liquidity and efficient price discovery). In Chap. 5, I examine the association between the advancement of liquidity and the onset of trading in Phase II of the EU-ETS. The liquidity effects of other relevant effects are also examined. Since power companies and other large entities are compliance traders in the EU-ETS, one would expect regular institutional trading patterns such as block trading. Therefore, in Chap. 4, we investigate the impact of these trade classes on the EU-ETS's largest trading platform, the ECX.

Notes

1. These sectors include electricity generators, mineral oil refineries, coke ovens, ferrous metals, glass, ceramic products and cement manufacturers to glass and pulp producers. Electricity generators are however, the leading CO_2 emitters. By Council decision, the aviation sector joined the EU-ETS (see Directive 2008/101/EC).
2. CO_2 certificates, emission allowances, emission permits are used interchangeably to refer to the legal right to pollute in this book.

References

Alberola, E., Chevallier, J., & Chèze, B. (2008). Price Drivers and Structural Breaks in European Carbon Prices 2005–2007. *Energy Policy, 36*, 787–797.

Albrecht, J., Verbeke, T., & Clerq, M. (2004). Informational Efficiency of the US SO_2 Permit Market. *Environmental Modelling & Software, 21*, 1471–1478.

Boemare, C., & Quirion, P. (2002). Implementing Greenhouse Gas Trading in Europe: Lessons from Economic Literature and International Experiences. *Ecological Economics, 43*, 213–230.

Böhringer, C., & Lange, A. (2005). On the Design of Optimal Grandfathering Schemes for Emission Allowances. *European Economic Review, 49*, 2041–2055.

Butzengeiger, S., Betz, R., & Bode, S. (2001). *Making GHG Emissions Trading Work – Crucial Issues in Designing National and International Emission Trading Systems*. Hamburg Institute of International Economics Discussion Paper 154, Hamburg.

Capoor, K., & Ambrosi, P. (2009). *State and Trends of the Carbon Markets, 2009*. The World Bank Report, Washington, DC.

Christiansen, A. C., & Arvanitakis, A. (2005). Price Determinants in the EU Emissions Trading Scheme. *Climate Policy, 5*, 15–30.

Daskalakis, G., & Markellos, R. N. (2008). Are the European Carbon Markets Efficient? *Review of Futures Markets, 17*, 103–128.

Daskalakis, G., Psychoyios, D., & Markellos, R. N. (2009). Modeling CO_2 Emission Allowance Prices and Derivatives: Evidence from the European Trading Scheme. *Journal of Banking & Finance, 33*, 1230–1241.

Daskalakis, G., Ibikunle, G., & Diaz-Rainey, I. (2011). The CO_2 Trading Market in Europe: A Financial Perspective. In A. Dorsman, W. Westerman, M. B. Karan, & Ö. Arslan (Eds.), *Financial Aspects in Energy: A European Perspective* (pp. 51–67). Berlin; Heidelberg: Springer.

Dodwell, C. (2005). *EU Emissions Trading Scheme: The Government Perspective.* Paper Presented at the Business & Investors' Climate Change Conference 2005: 31/10-01/11, 2005, London.

Fezzi, C., & Bunn, D. (2009). Structural Interactions of European Carbon Trading and Energy Prices. *The Journal of Energy Markets, 2*, 53–69.

Frino, A., Kruk, J., & Lepone, A. (2010). Liquidity and Transaction Costs in the European Carbon Futures Market. *Journal of Derivatives and Hedge Funds, 16*, 100–115.

Grubb, M., & Neuhoff, K. (2006). Allocation and Competitiveness in the EU Emissions Trading Scheme: Policy Overview. *Climate Policy, 6*, 7–30.

Hanemann, M. (2009). The Role of Emission Trading in Domestic Climate Policy. *The Energy Journal, 30*, 79–114.

Hawksworth, J., & Swinney, P. (2009). *Carbon Taxes vs Carbon Trading.* PriceWaterhouseCoopers Report, London.

Hintermann, B. (2010). Allowance Price Drivers in the First Phase of the EU ETS. *Journal of Environmental Economics and Management. 59*, 43–56.

Hobbs, B. F., Bushnell, J., & Wolak, F. A. (2010). Upstream vs. Downstream CO_2 Trading: A Comparison for the Electricity Context. *Energy Policy, 38*, 3632–3643.

Joskow, P. L., Schmalensee, R., & Bailey, E. M. (1998). The Market for Sulfur Dioxide Emissions. *The American Economic Review, 88*, 669–685.

Kara, M., Syri, S., Lehtilä, A., Helynen, S., Kekkonen, V., Ruska, M., et al. (2008). The Impacts of EU CO_2 Emissions Trading on Electricity Markets and Electricity Consumers in Finland. *Energy Economics, 30*, 193–211.

Kling, C., & Rubin, J. (1997). Bankable Permits for the Control of Environmental Pollution. *Journal of Public Economics, 64*, 101–115.

Koch, N. (2012). *Co-movements between Carbon, Energy and Financial Markets: A Multivariate GARCH Approach*. School of Business, Economics and Social Sciences, University of Hamburg Working Paper, Hamburg.

Kossoy, A., & Ambrosi, P. (2010). *State and Trends of the Carbon Markets, 2010*. The World Bank Report, Washington, DC.

Labatt, S., & White, R. R. (2007). *Carbon Finance: The Financial Implications of Climate Change*. New Jersey: John Wiley & Sons.

Linacre, N., Kossoy, A., & Ambrosi, P. (2011). *State and Trends of the Carbon Market 2011*. The World Bank Report, Washington, DC.

Linares, P., Santos, F. J., Ventosa, M., & Lapiedra, L. (2006). Impacts of the European Emission Trading Directive and Permit Assignment Methods on the Spanish Electricity Sector. *The Energy Journal, 27*, 79–98.

Mansanet-Bataller, M., Pardo, T., & Valor, E. (2007). CO_2 Prices, Energy and Weather. *The Energy Journal, 28*, 73–92.

Montgomery, W. D. (1972). Markets in Licenses and Efficient Pollution Control Programs. *Journal of Economic Theory, 5*, 395–418.

Neuhoff, K., Ferrario, F., Grubb, M., Gabbel, E., & Keats, K. (2006). Emission Projections 2008–2012 Versus NAPs II. *Climate Policy, 6*, 395–410.

Reneses, J., & Centeno, E. (2008). Impact of the Kyoto Protocol on the Iberian Electricity Market: A Scenario Analysis. *Energy Policy, 36*, 2376–2384.

Rubin, J. D. (1996). A Model of Intertemporal Emission Trading, Banking, and Borrowing. *Journal of Environmental Economics and Management, 31*, 269–286.

Schennach, S. M. (2000). The Economics of Pollution Permit Banking in the Context of Title IV of the 1990 Clean Air Act Amendments. *Journal of Environmental Economics and Management, 40*, 189–210.

Schleich, J., Ehrhart, K.-M., Hoppe, C., & Seifert, S. (2006). Banning Banking in EU Emissions Trading? *Energy Policy, 34*, 112–120.

Schmalensee, R., Joskow, P. L., Ellerman, A. D., Montero, J. P., & Bailey, E. M. (1998). An Interim Evaluation of Sulfur Dioxide Emissions Trading. *The Journal of Economic Perspectives, 12*, 53–68.

Sijm, J. P. M., Bakker, S. J. A., Chen, Y., Harmsen, H. W., & Lise, W. (2005). *CO_2 Price Dynamics: The Implications of EU Emissions Trading for the Price of Electricity*. Energy Research Centre of the Netherlands (ECN) Working Paper ECN-C-05-081, Amsterdam.

Sijm, J., Neuhoff, K., & Chen, Y. (2006). CO_2 Cost Pass-through and Windfall Profits in the Power Sector. *Climate Policy, 6*, 49–72.

Springer, U. (2003). The Market for Tradable GHG Permits under the Kyoto Protocol: A Survey of Model Studies. *Energy Economics, 25,* 527–551.

Stavins, R. N. (1998). What Can We Learn from the Grand Policy Experiment? Lessons from SO_2 Allowance Trading. *The Journal of Economic Perspectives, 12,* 69–88.

Svendsen, G. T., & Vesterdal, M. (2003). How to Design Greenhouse Gas Trading in the EU? *Energy Policy, 31,* 1531–1539.

Vesterdal, M., & Svendsen, G. T. (2004). How Should Greenhouse Gas Permits Be Allocated in the EU? *Energy Policy, 32,* 961–968.

3

Price Discovery and Trading After Hours on the ECX

3.1 Introduction

It has been more than 135 years since the *Walrasian* theory of general equilibrium (price formation) was first considered; yet price formation in financial markets remains a very active area in financial economics research. Technological and regulatory changes since the late 1980s have altered the way markets operate, spurring a series of new studies into price formation on trading platforms. A number of studies have thus examined issues surrounding price discovery in financial markets and its determinants (see among others Barclay et al. 1990; Chan et al. 1995; Easley et al. 1996, 1997; Flood et al. 1999). Aided by technology, the introduction of AHT has further altered the landscape. With the option to trade after official market closes comes additional opportunities to incorporate new information into instruments prices, even after the closing bell has been rung. For sophisticated trades, there is now no need to wait until the next day trading session before taking advantage of new information in the market. The seminal works of Barclay and Hendershott

---

The study described in this chapter is based on Ibikunle et al. (2013).

© The Author(s) 2018
G. Ibikunle, A. Gregoriou, *Carbon Markets*,
https://doi.org/10.1007/978-3-319-72847-6_3

(2003, 2004) have contributed significantly to our understanding of the evolution of price after markets close. Barclay and Hendershott (2003) investigate the AHT periods of before market opens (BMO) and after market closes (AMC) on the NASDAQ, creating the first comprehensive insight into how these two periods contribute to price formation in financial markets and their contrasting features. A number of other contributions were previously made to BMO price discovery through the analysis of non-executed orders and non-binding quotes prior to opening. Madhavan and Panchapagesan (2000) and Stoll and Whaley (1990) analyse contributions that activities of professional traders make to the opening price on the NYSE. Biais et al. (1999) and Davies (2003) investigate the effect of non-binding BMO orders on the Paris Bourse and the Toronto Stock Exchange, respectively; they show how that orders submitted during the BMO, though not binding, reflect learning in the markets. Ciccotello and Hatheway (2000) and Cao et al. (2000) also examine the price discovery process by means of non-binding market maker quotes.

More recently, He et al. (2009) investigate the efficiency of price discovery in a 24-hour US treasury market showing that the overnight trading period is a more important component of the treasury price discovery process than previously thought. This is a clear departure from the findings of Barclay and Hendershott (2003) on contributions of overnight trades to price discovery. In their analysis of AMC trading, price contributions and discovery after the release of firm earnings (during AHT), Jiang et al. (2012) seem to arrive at the same conclusion as He et al. (2009). Confirming the clout of AHT, they find BMO and AMC periods contribute 36 and 60% of price discovery, respectively, on earnings announcements days despite comparatively low trading volumes (see also Greene and Watts 1996). Macroeconomic announcements as well have been linked with exchange rate jumps by Andersen et al. (2003), Almeida et al. (1998) and Goodhart et al. (1993). Andersen et al. (2003) also report news impact asymmetry in that bad news has a stronger impact than good news.

This chapter's study on AHT and price discovery differs from the aforementioned works in one respect. The investigations reported here are conducted in a unique market, the exchange-traded permits market or to be more apt, the European carbon market. In this market, the motivation for AMC trading year-round is difficult to grasp since permits are submitted to authorities for compliance purposes only once a year. Although permits have value, they are 'created' as records by regulatory authorities with the mechanism to influence prices if the need arises. Moreover, for emissions markets, transaction costs are incurred at various levels of trading. Additional costs incurred on information, seeking agreeable terms on block trades and reporting compliance records to authorities do contribute heavily to transaction costs (see Cason and Gangadharan 2003, 2011; Gangadharan 2000; Stavins 1995). Additional elements of transaction costs can result in a loss of trading volumes when trades are made unappealing (Kerr and Máre 1998). These issues are not all pertinent to regular markets Grüll and Taschini (2011).

There is a growing body of work on price formation in the EU-ETS; most of these are based on Phase I data, a number of which have been discussed already in the introductory chapters. Other relevant studies are discussed in Chap. 6 of this book. These studies, do not investigate intraday evolution of price discovery or efficiency of price formation and their relationship to trading activity in carbon markets. These issues are addressed in this study using data based on exchange transactions during Phase II of the EU-ETS.

Fundamental shifts[1] possible in the composition of informed and uninformed traders during the transition from RTH to the AMC period[2] provides an opportunity to employ some of the techniques used in previous AHT studies in this analysis of the carbon permit market. This chapter is aimed at answering the following questions: First, what is the impact of trading activity on the disclosure of information in carbon futures and does this have any effect on the time of disclosure? Second, does the AMC period in carbon futures reveal anything? Third, does liquidity inform price discovery and information efficiency in carbon futures market? And fourth, what is the impact of trading activity on informational efficiency of carbon futures contract prices?

This book's major contribution to the existing literature on price discovery in AHT comes from this chapter's analysis of a unique market which exchanges tradable carbon permits. The EU-ETS is the world's first large experiment for CO_2-based emission trading, and policy makers in other regions throughout the world are keenly watching it. Its success or otherwise will be a factor in determining whether a global emissions trading scheme will be adopted as a mechanism for limiting GHG emissions. The findings here on intraday evolution of price discovery during the RTH and AHT period will contribute to our understanding of whether the EU-ETS market is efficient, and therefore if it can provide a platform for a global market-led approach to tackling global warming via the reduction of carbon emissions.

The contributions of this study go beyond adding to the literature on market microstructure. Much of the microstructure properties of the European carbon futures market are still being unravelled; the results can therefore inform trading patterns especially for compliance buyers of carbon instruments. Regulators can also gain insights into the effect of certain trade instruments used in the market and on the trading process. In addition, participants currently trade bilaterally in the AMC period on the ECX with little-documented knowledge on the informational risks associated with this; the study presented in this chapter sheds more light on the associated risks.

The analysis presented here mainly compares different intervals/periods of the normal trading day/RTH and the AMC period. The preponderance of market microstructure studies establishes that information asymmetry reduces over the course of the day (see Glosten and Harris 1988; Huang and Stoll 1997; Lin et al. 1995; Madhavan et al. 1997 and others). This study is predicated on the role played by information over the RTH and the AMC periods; hence, the first step taken is to analyse the information asymmetry levels across the periods using the Huang and Stoll (1997) spread decomposition model. Adverse selection costs and effective spreads for each of the periods are obtained using the model. The estimated parameters provide the basis to launch this chapter's investigations.

The results presented in this chapter reveal that more contracts are traded per minute in the AMC period than during the normal trading day. Using the Huang and Stoll (1997) spread decomposition model, the study further shows that higher levels of information asymmetry are present during the AMC period/hour than at any other interval per hour during the normal trading day; the normal trading day is, however, responsible for the highest share of price discovery at over 71%. This is not surprising since there are ten hours within the normal trading day and just one of AMC trading. The research also finds evidence that contribution to price discovery is a function of liquidity. Less liquid contracts prove the highest contributors to price discovery, even though they are largely informationally inefficient. The analysis of exchange-traded permits thus agrees for the most part with the existing literature in that a small amount of trading activity can generate disproportionate price discovery and that liquidity is associated with informational efficiency. Furthermore, as in other studies, the least traded instruments contribute the largest proportion of price discovery in the AMC. The aggregate findings suggest that a mandatory cap and trading scheme such as the EU-ETS is an efficient way of reducing carbon emissions. The efficiency of the EU-ETS could provide a basis for the introduction of a mandatory global market-led approach to reducing carbon emissions.

The remainder of this chapter is structured as follows. First, in Sect. 3.2, the trading environment on the ECX is discussed as background to the analysis. Section 3.3 discusses the sample selection and describes the data. Section 3.4 describes the econometric methods employed and also presents and discusses the results of the empirical analyses and Sect. 3.5 concludes.

3.2 The Trading Environment on the ECX

Trading in physically delivered EUA futures commenced in April 2005. The contracts are offered on a quarterly expiry cycle (March, June, September and December) up to June 2013. EUA futures contracts with annual (December) deliveries for 2013 until 2020 have also been introduced. The underlying for each ECX EUA contract is 1000 EUAs. The trading system is electronic and continuous and begins at 7:00 hours and ends at 17:00 hours, the United Kingdom's local time from Monday to Friday. The maturity date for the contracts is the last Monday of the traded month and physical settlement occurs three days after expiry. In 2010, EUA carbon permits accounted for more than 84% of global carbon market value. Of these, approximately 73% are traded as futures contracts (see Kossoy and Ambrosi 2010; Linacre et al. 2011). The ECX platform is the market leader in EU-ETS exchange-based carbon trading with more than 92% market share. This includes OTC trades registered on the platform to reduce counter-party risk. The December maturity contracts roughly represent about 76% of daily transactions on the platform and hence form the basis of this study's investigations. The global dominance of the ECX platform has attracted participants from beyond Europe. In 2009, about 15% of trade volume on the platform was from traders domiciled in the United States (Kossoy and Ambrosi 2010).

Carbon Financial Instruments (CFI) trading on the ECX platform is done electronically on the ICE platform. ICE Futures Europe platform can only be accessed by members and this is strictly for the purpose of placing orders for execution. All trading orders placed on the platform and their corresponding executions are anonymous. The electronically executed trades go through the so-called Trade Registration System (TRS) for account allocation.

For most major futures trading platforms operated by Intercontinental Exchange (ICE) Europe, owners of the ECX, there is a pre-open trading period of 15 minutes to allow for participants to place early orders in ahead of the trading day. Thus, the market opens for an initial period between 6:45 and 7:00 am. This 15-minute period on the ECX platform hardly records any executed orders. Over the ten-month period covered by the dataset used for this study (February–November 2009), only 12 trades with a combined contract volume of 700 were recorded on the exchange. Trading takes place between 7:00 and 17:00 hours on the ECX platform. Presumably, no officially sanctioned trade is allowed beyond this point; however, allowance is provided for the registering of Exchange for Physical (EFP) and Exchange for Swaps (EFS) trades. EFP and EFS trades are only permitted for EUA and CER Futures contracts. This can be reported using an electronic facility called the ICEBLOCK on the platform. The Exchange officially requests that these trades be registered up to 30 minutes after the close of official trading each trading day except for the day of expiration of the traded contract. Reporting of EFP/EFS trades on the exchange, however, occurs up until about 6:00 pm regularly on the days covered by the sample as acknowledged by ECX officials. The trades are essentially bilateral trades that in theory require no intermediary/broker-dealer to execute.

The trades are usually registered by the buyer who charges the trade to the seller. The seller then matches the trade by confirming it (Non-crossed Trade). It is possible for prices of AMC EFP and EFS trades to fall outside the high and low points in the RTH (or beyond the biggest price shift from the previous close's settlement price) since the prices are not revealed until after the close, but this requires approval from the ICE Futures Europe Compliance Department prior to registration. The maximum price deviation, however, is pegged at €1.00.

EFP/EFS trades provide hedging options when trading ECX EUA futures contracts, in just a single transaction. Specifically, the seller of emissions permits becomes the buyer of ECX futures contract and the buyer of the permits, the corresponding seller of ECX contracts. They also allow for the substitution of Over the Counter (OTC) swap positions with corresponding ECX contracts. For these trades, there are no obligatory market maker quotes during the AMC. Members (brokers) can execute these trades on behalf of clients on the ICE platform, and the brokers are still under a strict obligation to obtain the best deals on behalf of their customers as their fiduciary duty demands.

Spreads and consequently, trading costs during the AMC are expected to be larger than during the RTH. It is expected that the volume of reported trades during the AMC will be very low (daily average) in comparison to the RTH. However, since these trades are likely to be executed by professional traders aiming to balance their positions at the end of a trading day, the volume per trade should be higher than during the RTH. The absolute necessity to satisfy optimal portfolio balancing therefore trumps the disadvantage of low liquidity and possible higher trading costs.

3.3   Data

3.3.1   Sample Selection

Two data sets from the ECX platform are employed for the analysis in this chapter. The first, which is the main dataset employed, comprises of all intraday tick-by-tick ECX EUA Futures contracts trades on the ICE platform from February 2009 through to November 2009. The dataset contains date, timestamp, market identifier, product description, traded month, order identifier, trade initiation (bid/offer), traded price, quantity traded, parent identifier and trade type. This dataset is provided directly by ICE Data LLP, London for the purpose of this doctoral research. The dataset contains 15 CFIs available on the ECX (futures contracts and futures spreads).

The second dataset is the end of day (EOD) data for ECX Futures from February 2009 through November 2009. From here, we obtain the daily settlement price, daily low price, daily high price and daily first execution price. We also extract the daily volume (for all trade types) and daily weighted average price from the dataset. This dataset does not contain any records for futures contract spreads, which are present in the first dataset.

Eleven of the CFIs included in the datasets are eliminated after applying the following conditions:

1. Both datasets must provide trading records from February 2009 through November 2009 for the selected CFI.
2. As the analysis is based on the comparison of RTH to AMC, the CFI must be tradable during both periods in an EFP or EFS trade.
3. The CFI must also be traded for at least 20% of days during both periods between February and November 2009.

The ECX tick dataset includes the trade initiator identifier, hence trades could be identified as buyer- or seller-initiated trades. For the AMC trades, however, the challenge is that all trades are identified as buyer-initiated by default since the trades are usually registered by the buyer who then alleges it to a seller. The TRS records trades based on the order submission and if it is matched, the order submitted first is the initiating order and becomes the initiator. This is misleading in the case of the AMC trade registration; hence to overcome this challenge, the tick test is used for classifying AMC trades. For robustness, results based on tick test are compared with those based on actual exchange identifier codes for the normal trading day, the results are very similar. Trades occurring at a price greater than the prevailing trade midpoint are classified as buyer-initiated and those at a price lower than the prevailing midpoint as seller-initiated. If the current and the previous trades are of the same price, then I classify using the next previous trade. Analysis of the tick rule by Lee and Ready (1991) and Aitken and Frino (1996a) suggest the tick rule's accuracy to be in the excess of 90%, with accuracy levels as low as 74% in some instances. Finally, the 12 trades recorded before 7:00 hours London local time in the sample are excluded from the final dataset.

3.3.2 Sample Description

The ECX platform is the only exchange where official AMC trading is recorded in the EU-ETS and the trading is quite thin, averaging 2700 contracts over 60 trades per day during our sample period. The contracts in the sample represent 99.998% of total AMC trades recorded on the platform between February and November 2009. Trading in the EU-ETS in comparison to more established financial and commodities derivatives markets is very thin. AMC trades averaged about €37,000,000 in value per day or 14.58% of the daily total Euro value between February and November 2009. AMC recorded trading is limited to between 17:00 and 18:00 hours GMT. As noted by Porter and Weaver (1998), block trades on NASDAQ in the past were posted late (after-hours) after having been executed during the RTH. This does not hold for the sample as the trading system is electronic and the participants have real-time access to input their trades at any point during the RTH. Furthermore, the bilateral nature of the EFP and EFS trades ensures that the trades even though agreed during the trading are unknown to the market until it is registered on the ICE platform. Finally, the order Ids acquired by the trades evidenced that they all entered the system after the close of trading, which is as far as trade execution time can be approximated, moreover, an order on ICE platform is only executed when it is matched by a corresponding buy/sell order.

Table 3.1 is the summary of RTH and AMC trading activity. The results show daily average estimates for individual contracts and the full sample averages per contract per day. The trading volume is skewed towards one

Table 3.1 Trading summary

Futures contracts	AMC				RTH		
	Number of trades	Volume (€ '000)	Contract volume	% of days with trading	Number of trades	Volume (€ '000)	Contract volume
Dec-09	50.23	25,907	1957.07	97.22	1239.61	129,263	9775.74
Dec-10	5.66	4603	328.18	90.28	142.33	46,571	1674.24
Dec-11	1.49	1916	131.03	63.89	52.51	19,669	681.29
Dec-12	3.18	4214	267.79	79.63	84.65	19,769	1279.83
Dec-13	0.12	193	11.40	7.41	1.09	515	32.85

The table shows a summary of trading activities of AMC and RTH periods for five contracts trading on the ECX platform. The contracts are the highest volume trading ones of the contracts eligible for AMC trading. The data covers the trading period from February 2009 through to November 2009. The table includes estimates for daily average euro volume, number of contract trades executed per day, average contract volume per day and percentage of days with trading for the AMC period. The RTH period runs between 7:00:00 and 16:59:59 hours London time; the AMC runs between 17:00 and 18:00 hours London time

contract (Dec-2009) throughout the 211 days covered by this study. About 83% of the AMC trades in this sample are recorded for the December 2009 contract. This is not unusual in the EU-ETS; we report the same phenomenon for the Leipzig-based European Energy Exchange (EEX) in Chap. 5 of this book. Joyeux and Milunovich (2010) as well as Mizrach and Otsubo (2014) also report the same trend for the ECX at various time in Phases I and II, respectively.

The most-traded contract, the Dec-2009, averages 50 AMC trades per day (with a market value of about €26,000,000 per day), while the other four contracts in this sample account for an approximate average of ten trades per day. Trading activity for the lower trading contracts shows a steep fall to an average of about six contracts per day for the closest trading one (Dec-2010) to the Dec-2009 contract. The Dec-2013 contract has an average less than 0.12 trades recorded during the AMC period. As a result of this extremely low level of trading activity, the contract and the others with lower trading AMC activity are excluded from further analysis (with the exception of trading activity analyses in this section). The inclusion of the Dec-2013 contract gives a more accurate picture of trading activity but due to low level of transactions, it could not be included in more robust analyses carried as part of this chapter.

3.4 Results and Discussion

3.4.1 Trading Volume and Volatility

Figure 3.1 depicts for each half-hour interval, the average daily trading volume and average return volatility. Volatility is computed as the standard deviation of half-hour returns on the platform, it is estimated using the December 2009 contract only due to large gaps in the trading cycle of the other contracts. Trading on the ECX displays an inverted S-shape rather than the now familiar U-shape pattern identified with derivative markets (see Chan et al. 1995; Gwilym et al. 1997). Also, in a clear

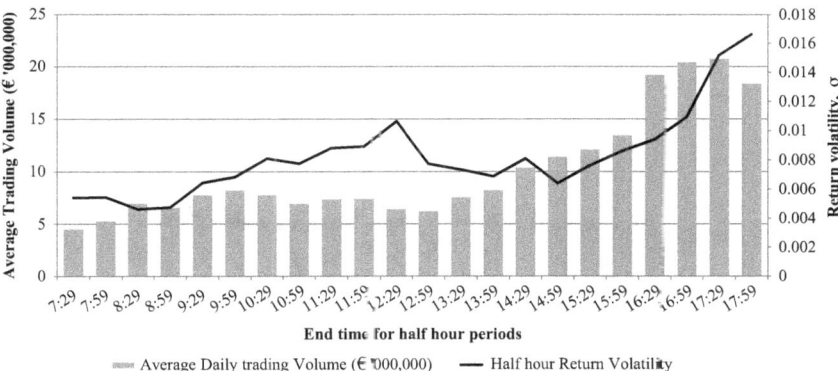

Fig. 3.1 Trading volume and volatility. The figure shows average daily trading volume and volatility for every half-hour period of the RTH and AMC periods for the December maturity contracts (2009–2013) trading on the ECX platform. The data covers the trading period February 2009 through November 2009. Volatility is computed as the standard deviation of the half-hour contract return and is calculated for the December 2009 contract only. Volatility is estimated using the December 2009 contract only due to significant gaps in the trading cycle of the other contracts. The RTH period runs between 7:00:00 and 16:59:59 hours London time; the AMC runs between 17:00 and 18:00 hours London time

contrast to previous studies, the most active period during the RTH is not the opening period. The closing stages of the RTH and the 1 hour AMC period are the most active periods on the ECX volume-wise. Indeed, the largest Euro volume of trades per half-hour is recorded in the first half-hour of AMC trading. This volume then holds steady for the concluding half-hour of the AMC trading period shedding only about 11.63% of volume from the preceding half-hour period. Volatility and trading volume show high levels of correlation (Spearman's rank correlation coefficient of 0.74) as illustrated in Fig. 3.1. The high level of correlation corresponds to the current evidence from most equity trading platforms, for example, Barclay and Hendershott (2003) document high correlation levels for NASDAQ.

While the AMC trading volume on the ECX platform is inconsistent with previous studies (see as an example Barclay and Hendershott 2003), larger AMC mean and median trade values are consistent with them. Figure 3.2 shows log-transformations of median and mean trade sizes at one-minute intervals, the log scale is used because of the large variability of the trade sizes after-hours. As expected, very steep rises in the mean

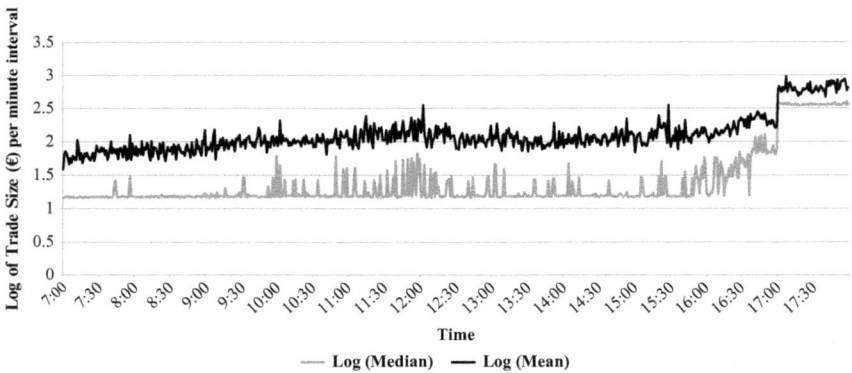

Fig. 3.2 Log median and mean trade size. The figure shows the logarithmic conversion of median and mean trade sizes over the entire trading periods of RTH and after market closes (AMC) for the December maturity contracts (2009, 2010, 2011, 2012 and 2013) trading on the ECX platform. The data covers the trading period February 2009 through November 2009. The RTH period runs between 7:00:00 and 16:59:59 hours London time; the AMC runs between 17:00 and 18:00 hours London time

and median sizes of trades are observed after the close. Mean estimates nearly quadruple in the first minute of AMC trading and then peak at almost €955,000 on 17:07 hours. Trading values during the remainder of the session hold up competently. The final minute of trading has a mean value of nearly €600,000 which is almost twice the highest mean estimate at any point during the RTH.

3.4.2 Motivation for Trading During the AMC: Liquidity or Information?

There are no significant differences between the rules governing the registration of a trade on the ECX during RTH and AMC periods. However, trading during the AMC is a privilege open only to members with access to the ICEBLOCK facility. Given this, it is expected that the market at that time will be primarily composed of professional traders executing proprietary orders or acting on behalf of their clients. The question then arises, why leave it that late? Why is the largest value of trading per minute reserved for this period of the day? EFP and EFS trades can be reported at any time of the day with less of the time constraint that is the one-hour AMC market.

Microstructure studies assume that there are two main types of traders, both holding idiosyncratic risks: the first are those who trade in search of liquidity, their aim is inventory control and portfolio rebalancing (especially at the end of each regular trading day). The second set of traders comprises of those trading on private information unknown at the time of trading to the rest of the market. Since participation in the market is motivated by different reasons, it is assumed that these two classes of traders will have varying degrees of activeness in the AMC session.

According to the literature, information asymmetry and uncertainty over fundamentals underlying instrument valuations usually decrease over the course of a trading day (see Easley et al. 1997; Easley and O'Hara 1992; Foster and Viswanathan 1990; Huang and Stoll 1997; Kyle 1985; Madhavan et al. 1997). Barclay and Hendershott (2003) note that public and private information accrue overnight when no trading takes place. Indeed, for the ECX, there is a non-trading interval of 12 hours and 45 minutes between the end of the AMC market and the 15-minute pre-open at 6:45 hours London local time. This interval is longer than the 11-hour 15-minute combined trading period afforded by the platform. It is a safe bet that some information akin to the carbon/emissions permit or ancillary markets could have accumulated during the no-trading period, hence a high level of information asymmetry is anticipated for the early trading period and this is also expected to decline from then on and over the course of the normal trading day.

Furthermore, the EU-ETS is an unusual market with many twists and turns, fuelled by the controversy that is climate change science. A lot of the trades are aimed at offsetting emissions by market participants; hence, these would be based on information regarding the anticipated level of emissions. Installation production levels determine emissions, so there is an expectation of a market heavy on informed trades especially in the AMC period. The expectation includes a higher level of informed trading for the later maturity contracts than in the Dec-2009 contract. Emission permits are not submitted year-round to offset verified emissions positions; therefore, most trades are more likely based on the need for risk hedging and portfolio diversification than for submission in lieu of emission reduction. If emission permits are for submission to the authorities once a year, it is logical to expect that AMC trading is not an absolute necessity unless there is an advantage to be gained. Uhrig-Homburg and Wagner (2009) argue along the same lines in a study investigating the EUA spot and futures relationship. Much more importantly, the size of the typical EFP/EFS trades in this sample suggests that these trades are being mainly initiated by institutional and compliance traders, who are clearing members of the exchange. Kurov (2008) shows evidence that institutional traders in index futures markets are more informed than other classes of market participants. These reasons lead one to expect that larger spreads and higher levels of informed trading would subsist during the AMC period. In addition, previous studies have shown that there is a higher level of informed trading during the AMC than during the RTH (see Barclay and Hendershott 2003).

For trades motivated by the search for liquidity, the motive contrasts that of those influenced by privately held information. According to Brock and Kleidon (1992), being in possession of a sub-prime portfolio overnight comes at higher costs in comparison with holding an optimal one. This is a big enough motivation for market participants who could not complete the optimal balancing of their portfolio during the regular hours to trade during the AMC period. On the ECX, the EFP and EFS trades offer this opportunity.

In addition to expecting higher informed trades during the AMC, it is also anticipated that the first few hours of the RTH will exhibit a higher level of information asymmetry than the other intervals during the RTH. This is as a result of the information accumulation we earlier noted that would have occurred during the non-trading period overnight. Although there is a higher CFI traded volume during the AMC period, there will be reduced liquidity, due to a reducing number of traders during this period of the day. Based on the foregoing, one can assume that the trading spreads during the AMC period would be larger than at any interval during the RTH period. The superior volume per trade in the AMC suggests that a substantial proportion of the trades are information induced (see Chan and Lakonishok 1993 and others; Easley and O'Hara 1987; Kraus and Stoll 1972), hence the larger spreads compensate market makers and uninformed traders for the risk of trading with informed traders.

This chapter thus investigates whether or not there is a higher level of informed trading and larger spreads during the AMC trading hour than at any other period during the regular trading day, by estimating adverse selection information components and effective spreads at different intervals during the regular trading day and during the AMC period. The approach used is based on estimating adverse selection costs for various intervals by using Huang and Stoll's (1997) spread decomposition model, since adverse selection cost is a component of the bid-ask spread. First, in Sect. 3.4.3, we initially explain the intuition behind the main forerunner to the Huang and Stoll (1997) model, the Madhavan et al. (1997) trade indicator model, and then we explain how the Huang and Stoll (1997) model reconciles it and previous models.

3.4.3 Adverse Selection Costs and Spread Analyses

Understanding the microstructure of financial markets is vital. The use of measures such as the bid-ask spread as gauges of liquidity plays an important part in this as the academic market microstructure literature evolves to harness its descriptive characteristics. Most papers investigating market liquidity (and to lesser extent market efficiency) have largely been focused on the bid-ask spread. Danielsson and Payne (2010) identify several reasons for this. The first being the growth of the asymmetric information and inventory costs literature in the 1980s (see as examples Glosten 1987; Glosten and Milgrom 1985; Ho and Stoll 1981, 1983; Roll 1984). Based on these studies, spread estimations became less ambivalent. Another reason is the progress recorded in development of robust estimators for spread components based on existing theories (see as examples Foster and Viswanathan 1993; Glosten and Harris 1988; Hasbrouck 1991a, b; Stoll 1989). Finally, more often than not, available microstructure databases contain only information on the spread, with little or incomplete information on other variables necessary in the measurement of liquidity.

Madhavan et al. (1997) build on previous studies to produce a structural model of intraday price formation that incorporates microstructure impacts and the revision of trading information shocks. Subsequently, the Huang and Stoll (1997) model (based on portfolio trading pressure) reconciles Madhavan et al. (1997) and preceding models to fully decompose the transaction spread into all of its component elements. Huang and Stoll (1997) provide two extensions of the basic microstructure model, one of which is based on portfolio trading pressure. This section provides a succinct derivation of the Madhavan et al. (1997) model as well as the Huang and Stoll (1997) model extension based on portfolio trading pressure.

3.4.3.1 Madhavan et al.'s (1997) Spread Decomposition Model

This model is an integration of several microstructure influences identified and evaluated in previous works (see Choi et al. 1988; Garbade and Silber 1979; Glosten and Milgrom 1985; Roll 1984; Stoll 1989). Consider that the trading environment for a risky instrument with time-dependent value is a kind of auction-quote driven market. Both the market maker and traders submitting limit orders can provide liquidity in this market. The liquidity providers using any of these mechanisms thus provide bid and ask prices at which they wish to execute transactions. Execution of orders within these quotes is considered plausible.

Denote P_t as the transaction price at time t for the stated risky asset with a constantly evolving value through time. Also, let x_t be a dummy variable indicating whether a trade is buyer- or seller-initiated (+1 if it is buyer-initiated and −1 if it is seller-initiated). In the event that a trade is regarded as both buyer- and seller-initiated, then x_t will take on the value of 0. Denote λ as the unconditional probability that a trade takes place within the quoted spread: $\lambda = \Pr[x_t = 0]$. In the same manner, let the assumption hold that buy and sell transactions can similarly occur such that $E[x_t] = 0$ and $\text{Var}[x_t] = (1 - \lambda)$.

New knowledge in the form of publicly available announcements bring about the revision of previously held conclusions by market participants; similarly, order-flow-induced stimulus can affect previously held beliefs. The announcements already in the public domain can cause the revision of previously held notions without any trading occurring. Let ε_t represent the evolution of beliefs owing to new information being released to the public domain between the interval $t-1$ and t, and let the assumption hold that ε_t is i.i.d. with $E[\varepsilon_t] = 0$ and $\text{Var}[\sigma_\varepsilon^2]$. Also, market makers are exposed to adverse selection costs, thus leading to occasional upward and downward adjustments in published bid and ask orders. The reconsideration of beliefs is deemed to be analogous to the uncorrelated order flow. The review of beliefs is said to hold a positively correlated relationship with innovation in the order flow (see Glosten and Milgrom 1985). The alteration of held beliefs resulting from the order flow can then be written as $\theta(x_t - E[x_t|x_{t-1}])$, where $(x_t - E[x_t|x_{t-1}])$ represents the shock in the order flow and $\theta \geq 0$ quantifies the permanent effect component of the 'order flow innovation'; this is the asymmetric information parameter. Substantial reviews of beliefs with regards to innovation in the order flow are indicated by large values of θ. Madhavan et al.'s (1997) assumption with respect to a fixed size of order is congruous with preceding works (e.g. see Choi et al. 1988; George et al. 1991; Glosten and Milgrom 1985; Roll 1984; Stoll 1989).

Let the expected estimate of an asset's value based on publicly available information and the trade type (buyer- or seller-initiated) be represented by μ_t. The change in previously held beliefs is then equal to the aggregate of adjustment in beliefs, which is as a result of publicly available information and innovations in the order flow, such that the expected value of the asset after a trade can be written as:

$$\mu_t = \mu_{t-1} + \theta\left(x_t - E\left[x_t|x_{t-1}\right]\right) + \varepsilon_t. \tag{3.1}$$

Both Madhavan et al. (1997) and Glosten and Milgrom (1985) agree that the bid price is determined by a transaction being seller-initiated and the ask price being buyer-initiated. Therefore, the pre-transaction ask price at time t is denoted as p_t^a and similarly the bid price is denoted as p_t^b. As previously stated, market makers make their quotes with the intention of obtaining compensation for supplying liquidity when required by the market. Denote $\varphi \geq 0$ as cost/unit of an asset to the market maker due to her providing liquidity. The market maker's recompense to cover the costs of transaction, inventory and adverse selection is then φ. Thus, the price dependent on the dummy variable x_t being +1 (the ask price) can be expressed as $p_t^a = \mu_{t-1} + \theta\left(1 - E\left[x_t|x_{t-1}\right]\right) + \varphi + \varepsilon_t$. Following after the determination of ask price, the bid price can be expressed as $p_t^b = \mu_{t-1} + \theta\left(1 + E\left[x_t|x_{t-1}\right]\right) + \varphi + \varepsilon_t$. The parameter φ consequently encapsulates the transient impact of order flow on prices. Orders are carried out outside and at bid and ask prices as well as within them; the general assumption that runs through here will be that the trades are executed at the mid-price: $\left(p_t^a + p_t^b\right)/2$. The trade price is then computed as:

$$p_t = \mu_t + \varphi x_t + \xi_t \tag{3.2}$$

ξ_t is an *i.i.d.* random variable and has a mean of zero. ξ_t encapsulates the impact of randomly determined rounding errors due to price disjunction or variability of returns. It should be noted that due to a tendency to round up on buys and down on sells, there is likely to be a structured divergence of ξ_t from 0. The market maker cost component, φ captures this divergence.

It should also be noted that in Eq. (3.2), μ_t is the mean conditioned on the observance of x_t. Notwithstanding the fact that bid and ask prices from market makers are set prior to the commencement or conclusion of the trades, Madhavan et al. (1997) reason that these prices are still conditioned on the yet to be observed transaction signal. Buy orders are effected at the ask price and sell orders at the bid price. In fixing these prices, the market maker is compelled to include the cost associated with the trade, that is, φ, into the quoted prices. As a result of this inference, the transaction signal is already incorporated in the order execution/transaction price at time t. Now, employing the two Eqs. (3.1) and (3.2), one can infer that:

$$p_t = \mu_{t-1} + \theta\left(x_t - E\left[x_t | x_{t-1}\right]\right) + \varphi x_t + \varepsilon_t + \xi_t. \qquad (3.3)$$

Equation (3.3) can then be estimated by characterising the temporal functioning of the order flow. Madhavan et al. (1997) assume a general Markov chain for the trade initiation variable. γ denotes the probability distribution that a trade executed at a bid price succeeds a trade executed at an ask price and vice versa. This can be expressed as $\gamma = \Pr[x_t = x_{t-1} | x_{t-1} \neq 0]$. For ease of execution, block traders usually split their orders into smaller sizes. This translates into the assumption that orders are more likely to be sustained following an expression of interest rather than after a reversal of orders. Hence, $\gamma > 1/2$. Other factors related to exchange trading mechanism and so on can also be responsible for this practice.

The first-order autocorrelation of the transaction initiation variable is next denoted by ρ. This can be expressed as $\rho = \dfrac{E[x_t\, x_{t-1}]}{\text{Var}[x_{t-1}]}$. It can then be shown that $\rho = 2\gamma - (1 - \lambda)$, such that the autocorrelation of the order flow component, ρ, is an expanding function of γ and λ. When the likelihood of trading within the quotes (λ) is 0 and $\gamma = 1/2$ (i.e. γ is independent), the order flow is uncorrelated ($\rho = 0$).

Madhavan et al. (1997) give a transformation of Eq. (3.3) into a form in which it can be used in decomposing spread components. The authors substitute out the belief lag one period (the prior belief), μ_{t-1}. Contingent on $\mu_{t-1} = p_{t-1} - \varphi x_{t-1} - \xi_{t-1}$ and $E[x_t | x_{t-1}] = \rho x_{t-1}$, Eq. (3.3) can therefore be written as:

$$p_t - p_{t-1} = (\varphi + \theta)x_t - (\varphi + \rho\theta)x_{t-1} + \varepsilon_t + \xi_t - \xi_{t-1}. \quad (3.4)$$

When friction exists in the market, in the application of Eq. (3.4), the trading price shifts will evidence the order flow due to market inefficiencies and public information impacts. When there are no frictions in the market, however, Eq. (3.4) becomes representative of an efficient market following the classical random walk.

3.4.3.2 Estimating Madhavan et al.'s (1997) Spread Decomposition Model

Based on the Madhavan et al. (1997) model, one can conclude that the functioning of the market in terms of trade prices and quotes are subject to four parameters: the asymmetry information parameter, θ; the cost of liquidity provision by the market maker parameter, φ; the probability of a trade occurring within the quotes parameter, λ and the autocorrelation of the order flow parameter, ρ.

According to Madhavan et al. (1997), the parameter vector $\beta = (\theta, \lambda, \gamma, \rho)$ can be estimated using the generalised method of moments (GMM) estimator standardised by Hansen (1982). The GMM is a more suitable option than other estimation methods such as maximum likelihood as robust hypotheses for the stochastic flow generating the data is not needed. It also permits adaptations for structures of autocorrelation and conditional heteroscedasticity (see Huang and Stoll 1994, 1997; Madhavan et al. 1997). The model has five moment conditions specifically identifying the parameter vector β and the drift term (constant) α:

$$E \begin{pmatrix} x_t x_{t-1} - x_t^2 \rho \\ |x_t| - (1-\lambda) \\ \mu_t - \alpha \\ (\mu_t - \alpha)x_t \\ (\mu_t - \alpha)x_{t-1} \end{pmatrix} = 0. \quad (3.5)$$

The first condition specifies the first-order autocorrelation in quotes; the second is a definition of the probability of a trade occurring at mid-quote, the third-moment condition is a definition of the expectation of the drift term as the mean asset pricing error. The final two-moment conditions follow the normal form of OLS estimation. Under a GMM estimation, the parameter vector is computed to ensure the sampling of the population moments is executed such that the sample best estimates the population based on a specific weighting matrix, the choice of which is irrelevant. The asymptotic normality and consistency of the GMM parameter vector estimates are validated by Hansen (1982).

3.4.3.3 Huang and Stoll's (1997) Three-Way Spread Decomposition Model Based on Portfolio Trading Pressure

This approach employs the fact that quote shifts due to inventory costs do not arise from inventory alterations in just one instrument (i.e. the instrument of interest) but from other instruments held in a portfolio along with the instrument we are interested in. This is a portfolio approach to decomposing the spread; it is based on the assumption that adverse information relates to instruments on an individual basis, but inventory impacts are portfolio wide. In employing this approach, we assume that 'liquidity suppliers' execute ECX trades in the sample. This is not to be interpreted as meaning trades by market makers, but as trades executed by participants whose actions provide availability of required orders in the market. These orders help enhance market liquidity and since they are limit orders submitted to a limit order book, they also impose the pricing restrictions akin to a market maker. This is a simple view of the market that has been adopted by some microstructure studies (e.g. see Madhavan et al. 1997). Indeed Huang and Stoll (1997) propose a refinement of their approach through the nomination of specific portfolios other than a market maker's, for example, a specialist portfolio or traders using limit orders (as is the case with the data employed in this chapter's investigations).

Consider a liquidity supplier purchasing instrument x at the bid quote, the trade will lower the bid and offer prices of the instrument as well as other correlated instruments; and the sale in the correlated instruments, hedges his position in instrument x. In reverse, holding the assumption that the other instruments are constrained by trading pressure, the liquidity supplier can choose not to induce the lowering of the quoted prices for instrument x if his aim is to hedge his buying of other instruments thereby spurring sales in instrument x. This approach recognises that there is a probability that instrument x's quotes are driven by more than the inventory impacts and information components of only instrument x. Specifically, trading pressure on account of other instruments should result in alterations in quotes of instrument x due to the efforts of liquidity suppliers to retain the balance of their portfolios.

3.4.3.4 A Simple Model

Assuming no transaction costs, V_t the hidden core value of an instrument is established just before the bid and offer quotes are published at time t. Quote mid-price denoted as M_t will only be computed as the quotes are released. Let trade price at time t be P_t and Q_t, the purchase/sale indicator variable for the trade price, P_t. Q_t equals +1 if the trade is initiated by the buyer and also executes at a price higher than the mid-price, −1 if the trade is seller-initiated and also executes below the mid-price and finally takes the value of 0 if the trade executes at the mid-price. The hidden V_t is modelled as below:

$$V_t = V_{t-1} + \alpha \frac{S}{2} Q_{t-1} + \varepsilon_t, \qquad (3.6)$$

S corresponds to the constant spread, α is the percentage of the half-spread due to adverse selection costs, while ε_t represents the serially uncorrelated public information shock. Equation (3.6) decomposes transaction costs, V_t into two elements. The first is the private information element uncovered from the previous trade, $\alpha \frac{S}{2} Q_{t-1}$ (see Copeland

and Galai 1983; Glosten and Milgrom 1985). And the second is the public information element encapsulated by ε_t. Although, the transaction cost V_t is purely theoretical, the midpoint, M_t of the spread is observable. Suppliers of liquidity aim to achieve inventory equilibrium; therefore, they effect equilibrium-inducing transactions by modifying quotes (hence the mid-price and its evolution). The adjustments are carried out in relation to the core value of traded instruments as informed by their inventory levels (e.g. see Ho and Stoll 1981; Stoll 1978). Suppose that previous transactions are of a regular size of one, the midpoint (mid-price) in relation to the core instrument value then corresponds to:

$$M_t = V_t + \beta \frac{S}{2} \sum_{t=1}^{t-1} Qi, \qquad (3.7)$$

β corresponds to the magnitude of the half-spread measure due to inventory costs, where $\sum_{t=1}^{t-1} Qi$ is the aggregate inventory from when the market opens to time $t-1$, and Q_1 is the inceptive inventory for that trading day. If there are no inventory holding costs, the ratio of V_t to M_t will be one. Equation (3.7) holds for the bid, offer and mid-prices since it is already assumed that the spread is constant.

First, differencing of Eqs. (3.7) and (3.6) suggests that quotes are generally modified to account for the inventory costs and the information exposed by the last transaction. Specifically, we have:

$$\Delta M_t = (\alpha + \beta) \frac{S}{2} Q_{t-1} + \varepsilon_t \qquad (3.8)$$

with Δ as the first difference operator.

Equation (3.9) cites the assumption of a constant spread:

$$P_t = M_t + \frac{S}{2} Q_t + \eta_t, \qquad (3.9)$$

where η_t is the error term and it encapsulates the deviation of the observable half-spread, $P_t - M_t$, from the constant half-spread, $\dfrac{S}{2}$, along with the inclusion of price discreteness induced rounding errors. The estimable traded spread, S, is distinguishable from the observable spread, S_t, because it is representative of trades outside the midpoint but within the spread. Transactions within the quoted spread and that are executed above the midpoint are regarded as ask transactions, and transactions within the spread and that are executed below the midpoint are the bid transactions. When S is estimated, it will be larger than the effective spread, which is the absolute value of transaction price minus the prevailing midpoint, $|P_t - M_t|$. This is due to the exclusion of midpoint trades ($Q_t = 0$) from the estimation. In contradiction to this, the unobserved estimated spread S, obtained from serial covariance of transaction prices (see Roll 1984) are swayed by the volume of midpoint transactions, although Harris (1990) suggests that the Roll (1984) spread estimator may be significantly biased.

When Eqs. (3.8) and (3.9) are integrated, the basic regression model (3.10) is produced:

$$\Delta P_t = \frac{S}{2}(Q_t - Q_{t-1}) + \lambda \frac{S}{2} Q_{t-1} + e_t, \qquad (3.10)$$

whereby $\lambda = \alpha + \beta$ and $e_t = \varepsilon_t + \Delta\eta_t$. The regression model (3.10) is a nonlinear indicator variable model with within-equation restrictions. The requirement for estimation is the indication of whether the transactions at t and $t-1$ execute at any of ask, bid or mid-prices. The model estimation yields the traded spread, S, and the aggregate modification of quotes to transactions, $\lambda\left(\dfrac{S}{2}\right)$. The estimation Eq. (3.10) does not yield independent estimates of adverse selection component, α, and the inventory holding component, β. Nevertheless, the proportion of the half-spread which is not attributable to adverse information or inventory holding can be estimated as $1 - \lambda$. This is the order processing costs estimate.

3.4.3.5 Extension of Model Based on Portfolio Trading Pressure

Equation (3.10) does not consider the effects of normal trading pressure because the inventory modification earlier modelled in Eq. (3.7) is based on the individual instrument inventory held. The next step is therefore to differentiate transaction signs of the distinct instruments. If k corresponds to instrument k, Eq. (3.7) then becomes:

$$M_{k,t} = V_{k,\cdot} + \beta_k \frac{S_k}{2} \sum_{t=1}^{t-1} Q_{Ai}, \qquad (3.11)$$

where $Q_{A,t}$ is the aggregate buy-sell indicator variable defined as follows:

$$Q_{A,t-1} = 1 \text{ for } \sum_{k=1}^{n} Q_{k,t-1} > 0,$$

$$Q_{A,t-1} = -1 \text{ for } \sum_{k=1}^{n} Q_{k,t-1} < 0 \qquad (3.12)$$

$$Q_{A,t-1} = 0 \text{ for } \sum_{k=1}^{n} Q_{k,t-1} = 0,$$

where n is the number of instruments that suppliers of liquidity review to determine the mood of the market. Equation (3.11) can thus be written as

$$\Delta P_{k,t} = \frac{S_k}{2} \Delta Q_{k,t} + \alpha_k \frac{S_k}{2} Q_{k,t-1} + \beta_k \frac{S_k}{2} Q_{A,t-1} + e_{k,t}. \qquad (3.13)$$

The model (3.13) remains an indicator variable model, it reverts to Eq. (3.10) in the absence of portfolio trading effects/frictions. A key distinction, however, is that with Eq. (3.13) all spread components of the bid-ask spreads can be estimated individually.

Model (3.13) can be estimated using a GMM approach with appropriate adjustments to the orthogonality conditions. GMM levies relatively weak distributional requirements unlike maximum likelihood[3] (see Madhavan et al. 1997; Huang and Stoll 1997). In addition, it is necessary to align the trading times across all instruments involved in the estimation. In this section, we provide an analysis of the procedure adopted in order to ensure this alignment (see also Huang and Stoll 1997).

An econometric improvement controlling for correlations across instruments can be made to Eq. (3.13). As all instruments react to information in the public domain, the public information effects element in $e_{k,t}$ may be contemporaneously correlated across the instruments. A panel estimation of Eq. (3.13) may be more efficient according to the authors; however, the panel estimation forces a reduction in a number of observations and the inferences made with the time series estimations are essentially the same with the panel estimation inferences.

Since this approach basically models market participants as adopting a portfolio view when executing inventory modifications of stocks, it is related also to the Ho and Stoll (1983) model that shows the connection between quote shifts in a stock and contemporaneous shifts in the others. The authors show that the quote shifts in stock a which is in reaction to a transaction in stock b is contingent on $cov(R_a, R_b)/\sigma^2(R_b)$. The model has been established by several other studies (see Heflin and Shaw 2000; Van Ness et al. 2001). Van Ness et al. (2001) suggest that the Huang and Stoll (1997) model is superior to other commonly used models in measuring adverse selection costs. However, the 'superiority' of the model has its costs. Some authors have reported the possibility of obtaining implausible estimates from the model estimation when using the probability of trade reversal approach in place of trading pressure approach. For example, Clarke and Shastri (2000) report this problem analysing a sample of 320 NYSE firms, indeed Van Ness et al. (2001) also report similar issues. It seems that there is a correlation between a low probability of trade reversal and the implausible estimates. For this chapter, we report only the trade aggregator estimation and there is no evidence of this problem, especially since the estimates are comparable to those of Benz and Hengelbrock (2009) estimated using the Madhavan et al. (1997) model.

Equation (3.13) can be expressed simply as:

$$\Delta P_{k,t} = \beta_{1,k} Q_{k,t} + \beta_{2,k} Q_{k,t-1} + \beta_{3,k} Q_{A,t-1} + e_t, \quad (3.14)$$

where $\Delta P_{k,t}$ is the change in price from the previous retained trade, $Q_{k,t}$ is equal to 1 (-1) when the transaction at period t for contract k was a liquidity provider sell (buy) and $Q_{A,t-1}$ is the aggregate buy-sell indicator variable used in encapsulating portfolio trading pressure on market participants inventory levels. It is measured as in Eq. (3.12). The adverse selection cost component and the half-spread are thus computed by estimating Eq. (3.14) using the ordinary least squares as was adopted by Heflin and Shaw (2000).

This study follows Huang and Stoll (1997) in employing only the last trade at every five-minute interval in formulating the variables in Eq. (3.14).[4] Huang and Stoll (1997) observe that large trades are sometimes broken up and registered as smaller trades (see also Barclay and Warner 1993; Chakravarty 2001). In order to counter the problems that may arise from this, they employ a 'bunching' technique whereby trades occurring within five-minute intervals of each other, executed at the same price and with the same quotes are bunched together and regarded as one trade. They, however, point out that using one trade every five minutes as adopted in this study greatly reduces any problem that may arise from large trades being broken up and reported as smaller trades. Heflin and Shaw (2000) adopt this approach as well. Moreover, the results obtained by Huang and Stoll (1997) from the bunching technique suggest that the method increases the adverse selection component estimates. As devised in Eq. (3.14), the $\beta_{1,k}$ estimate is one-half the estimated effective spread, and the adverse selection component is equivalent to $2(\beta_{2,k} + \beta_{1,k})$. The Wilcoxon–Mann–Whitney test is used for obtaining statistical inference on the level of differences between the RTH intervals and the AMC period.

In Panel A of Table 3.2, the estimated adverse selection costs components of the effective spread for each contract and time interval are reported. The combined contracts' averages for the intervals are also reported. The results largely support the hypothesis on reducing information asymmetry over the course of RTH. In the RTH, information asymmetry is highest in earlier intervals. Overall, it is highest between 7:00

Table 3.2 Information asymmetry and half-spread by time interval

	Time periods						
	Normal trading day						AMC
	07:00–09:00	09:00–11:00	11:00–13:00	13:00–15:00	15:00–17:00	07:00–17:00	17:00–18:00
Panel A: Adverse selection costs							
Contracts							
Dec-2009	0.016	0.036	0.012	0.001	0.002	0.014	0.411
Dec-2010	0.044	0.024	0.024	0.032	0.000	0.022	0.403
Dec-2011	0.049	0.083	0.047	0.028	0.032	0.053	0.511
Dec-2012	0.084	0.062	0.058	0.080	0.036	0.063	0.489
Overall	0.048[a]	0.051[a]	0.035[a]	0.035[a]	0.017[a]	0.038[a]	0.453
Panel B: One-half-spread estimates							
Contracts	7:00–9:00	9:00–11:00	11:00–13:00	13:00–15:00	15:00–17:00	07:00–17:00	17:00–18:00
Dec-2009	0.026[b]	0.023[b]	0.025[b]	0.017[b]	0.017[b]	0.021[b]	0.235[b]
Dec-2010	0.045[b]	0.025[b]	0.032[b]	0.020[b]	0.018[b]	0.027[b]	0.239[b]
Dec-2011	0.050[b]	0.057[b]	0.042[b]	0.038[b]	0.025[b]	0.041[b]	0.324[b]
Dec-2012	0.072[b]	0.051[b]	0.083[b]	0.057[b]	0.038[b]	0.056[b]	0.348[b]
Overall	0.048[a]	0.039[a]	0.045[a]	0.033[a]	0.024[a]	0.036[a]	0.286

The table shows adverse selection costs components in Panel A and one-half-effective spreads in Panel B for the four highest volume December maturity contracts on the ECX platform. Both the adverse selection costs components and the one-half-spread components are estimated using the following contract specific model (Huang and Stoll 1997) using ordinary least squares with Newey and West (1987) HAC:

$$\Delta P_{k,t} = \beta_{1,k} Q_{k,t} + \beta_{2,k} Q_{k,t-1} + \beta_{3,k} Q_{A,t-1} + e_t,$$

where $\Delta P_{k,t}$ is the change in price from the previous retained trade, $Q_{k,t}$ is equal to 1 (−1) when the transaction at period t for contract c was a sell (buy) and $Q_{A,t-1}$ is the aggregate buy-sell indicator variable used in encapsulating portfolio trading pressure on market participants inventory levels, it equals 1(−1, 0) when the sum of $Q_{k,t-1}$ across all four contracts is positive (negative, zero). Adverse selection costs component for each interval in Panel A is given as:

$$2(\beta_{2,k} + \beta_{1,k})$$

One-half-effective spread for each interval in Panel B is given as $\beta_{2,k}$. Pairwise Wilcoxon–Mann–Whitney U tests are used to compute p values for the differences between each of the different contract-dependent normal trading day intervals and the AMC period

[a]Normal trading day intervals during which the contract estimates are significantly different from the AMC
[b]Statistical significance of the spread estimates at 1% level. The data covers the trading period February 2009 through November 2009. The normal trading day period runs between 7:00:00 and 16:59:59 hours London time, and the AMC runs between 17:00 and 18:00 hours London time

and 11:00 hours than at any interval during the rest of RTH. As expected there is a high level of information asymmetry after the close to support the suggestion that those who trade in this market do so based on private information. Following the explanation in Sect. 3.4.2, the uniqueness of this market lends credence to the conjecture that this is the case. The results are also consistent with earlier studies finding higher levels of informed trading in the AMC period than during the RTH (see Barclay and Hendershott 2003; He et al. 2009; Jiang et al. 2012). The average adverse selection cost component for all contracts during the AMC is almost 12 times the value for the normal trading day (07:00–17:00). This implies that participants are significantly more likely to trade with private information in the AMC market environment than during the normal trading day. Although the investigations on information asymmetry is conducted in a relatively less active and quite unusual market, the adverse selection cost and half-effective spread estimates in the RTH period are comparable to estimates from most previous studies (see as examples George et al. 1991; Glosten and Harris 1988; Heflin and Shaw 2000; Huang and Stoll 1997; Lin et al. 1995; Madhavan et al. 1997).

Panel B presents the effective half-spread estimates. The results confirm the hypothesis that spreads are wider in the AMC period than in the RTH. The results in the RTH are also comparable to the results obtained by Benz and Hengelbrock (2009) using the Madhavan et al. (1997) model to estimate half-spread in the ECX during Phase I of the EU-ETS. Spreads are generally higher during the first 2 hours of trading than at any other period during the RTH. All the half-effective spread estimates are statistically significant.

3.4.4 Price Discovery and Information Absorption on the ECX

It is established in the market microstructure literature that price discovery is a function of trading activity (see French and Roll 1986; Jiang et al. 2012; Kim et al. 1999; Pascual et al. 2004). In Sect. 3.4.3, it is demonstrated that information asymmetry is higher during AMC session than during the normal trading day. One can also see that spreads grow larger in the AMC period as well. This implies that there should be a higher proportion of price discovery per hour taking place during the AMC. Results obtained during the open-close period by Barclay and Hendershott (2003) suggest that the least trading instruments contribute the higher proportion of price discovery during the RTH. Table 3.1 shows that the December-2011 and December-2012 contracts are the least trading; therefore, the expectation is that they will contribute the highest ratio of price discovery during the normal trading period. The study now turns to examining how the level of trading activity influences the incorporation of information during the two periods (and the various intervals). We also examine the relatedness of the distribution of informed trading to price discovery by first computing information contribution ratios across time; we use two price contribution measures: the weighted price contribution (WPC) and the weighted price contribution per trade (WPCT).

3.4.4.1 Weighted Price Contribution

The WPC measure, which is already established by previous studies (see Barclay and Hendershott 2004; Barclay et al. 1990; Barclay and Warner 1993; Cao et al. 2000; van Bommel 2011),[5] is adopted as a measure of price discovery. The WPC of the EUA futures contracts during five intervals of the RTH and the one-hour AMC period are calculated. Also, derived are estimates for the total price discovery for the RTH period (07:00–17:00 hours London local time). The terminal period for the RTH is the last trade at or before 17:00:00 hours and the AMC period as the first trade after 17:00:00 hours. The WPC measure used estimates the proportion of the 24 hour (close-to-close) EUA contract price return that takes place at that period.

For each contract, this study defines the WPC for each 24 hour period and each period k as:

Table 3.3 Weighted price contribution by time intervals

	Time periods							
	Normal trading day						AMC	
Contracts	07:00–09:00	09:00–11:00	11:00–13:00	13:00–15:00	15:00–17:00	07:00–17:00	17:00–18:00	Days with zero price change
Dec-2009	−0.006[a]	−0.006	−0.02	−0.034	0.038	−0.028	0.003	0.019
Dec-2010	0.083[a,b]	0.038[a]	−0.027	0.04	0.059[a,b]	0.193[a,b]	−0.028	0.019
Dec-2011	0.051	0.015[a]	0.037[a]	0.042	0.084[b]	0.229[b]	0.081[b]	0.024
Dec-2012	0.084[b]	0.072[b]	0.029[a]	0.026[a]	0.109[b]	0.32[b]	0.23[b]	0.019
Overall	0.212[b]	0.119	0.019	0.074	0.29[a,b]	0.714[a,b]	0.286[b]	0.019

The table shows WPC of six normal trading day intervals and the AMC period to the close-to-close return for the four highest volume December maturity contracts on the ECX platform. For each contract and interval k the WPC is computed for each day and then averaged across days:

$$WPC_{k,c} = \left| \frac{|ret_c|}{\sum_{c=1}^{C}|ret_c|} \right| \times \left(\frac{ret_{k,c}}{ret_c} \right)$$

where $ret_{k,c}$ is the log-return for interval k and for EUA contract c. ret_c is the close-to-close return for contract c. The trading days when close-to-close returns equal 0 are eliminated. The final column shows the fraction of days with close-to-close return equal to 0. The overall estimate in the final row is the sum of WPC for all contracts in that time interval. Wilcoxon–Mann–Whitney (tie-adjusted) tests are used to determine whether contract-dependent values for normal trading day intervals are significantly different from the AMC period

[a] Contract-dependent normal trading day interval during which the contract WPC is significantly different from that of the AMC
[b] WPC values significantly different from 0 at the 5% level. The data covers the trading period February 2009 through November 2009. The normal trading day period runs between 7:00:00 and 16:59:59 hours London time; the AMC runs between 17:00 and 18:00 hours London time

$$WPC_{k,c} = \left(\frac{|ret_c|}{\sum_{c=1}^{C}|ret_c|}\right) \times \left(\frac{ret_{k,c}}{ret_c}\right), \qquad (3.15)$$

where ret_c is the close-to-close return for contract c and $ret_{k,c}$ is the log-return for period k and for EUA contract c. The intuition behind the WPC is that $\frac{ret_{k,c}}{ret_c}$ is measure of relative proportion of the day's return provided by contract c and $\frac{|ret_c|}{\sum_{c=1}^{C}|ret_c|}$, the standardised absolute value of ret_c, is the weighing factor for each contract. It ensures that values with smaller $|ret_c|$ are given small weight. Thus, the WPC is computed for each contract and average across days to obtain the WPC for each time period for each contract. The WPC across all the contracts is also reported. This is defined as:

$$WPC_k = \sum_{c=1}^{C}\left(\frac{|ret_c|}{\sum_{c=1}^{C}|ret_c|}\right) \times \left(\frac{ret_{k,c}}{ret_c}\right). \qquad (3.16)$$

Normally, the WPC is computed instrument by instrument and then averaged out across the instruments (see Cao et al. 2000). When this is the case, however, instrument correlations generated by the common constituent in the returns make statistical inferences very complex when using the mean WPC. Since we report the WPC individually for each contract, we are not concerned about this; therefore, the standard t-statistic is applied to test the null that the daily WPC values (per period and for each contract) are not significantly different from zero. The Wilcoxon–Mann–Whitney test is also used for obtaining statistical inference on the level of differences between the RTH intervals and the AMC period.

Table 3.3 shows the WPC estimates for 24 hours (close-to-close). The results show that the most liquid contract (Dec-09) contributes the least to price discovery over the entire trading periods. Recalling the results from Table 3.1, this suggests that price discovery contribution is a function of trading activity levels. Based on this, it is anticipated that the Dec-2011 and Dec-2012 contracts will be the highest contributors to price discovery. Results in Table 3.3 support this expectation: the Dec-2012 and the Dec-2011 contracts are the two highest contributors to price discovery over all the trading periods (55 and 31%, respectively). Together they account for 86% of total price discovery over the entire periods. Their contributions over the combined RTH period (07:00–17:00 hours) and the AMC period are statistically significant. This is to be expected because EU-ETS trading, as explained in Sect. 3.4.2, is dependent on information relating to emission levels, political and regulatory shifts on environmental legislations and global treaties. In this context, it is likely that the primary motivation for taking a position on a contract with maturity about three years away is the possession of information that this is a good move either for hedging or speculation.

Overall, most of the price discovery takes place during the RTH over a 10-hour period. However, more than a quarter of the price discovery occurs during the space of just 1 hour (17:00–18:00 hours) in the AMC trading period, despite the reduced number of executed trades. Another observation is that more than 21% of the total close-to-close price discovery occurs in the first 2 hours (07:00–09:00 hours) of the RTH. This is interesting considering the fact that only about 18.90% volume of trades for all periods occur during this period (see Fig. 3.1). Moreover, more than 88.13% of these trades are in the Dec-2009 contract that contributes very little to price discovery during the period. Effectively, only about 16.65% of the 62,872 trades taking place at this time in our sample hold significant price information. The consistency of this result with the results in Table 3.2, showing information asymmetry (for RTH) in general decline from the high levels of the morning trading period, is important. The hypothesis that there is an accumulation of information during the non-trading 12.75-hour period therefore holds. If this is the case, it is expected that individual trades in the opening period will contribute more to price discovery than the trades at any other period during RTH. The expectation here does not include the AMC because, although the period enjoys the highest Euro volume per minute of trade, the aggregate number of trades is vastly inferior to those in the RTH. This implies that the trades in the AMC are potentially as informative as the opening period. This hypothesis is tested in Sect. 3.4.4.2. It is also observed that the lowest average WPC during the periods is recorded during the 11:00–13:00 hours range. In Fig. 3.1, this period has the highest level of volatility relative to trading volume. The high return volatility and low WPC estimates raise the suggestion of a disproportionate level of noise in the price discovery process during the RTH. In Sect. 3.4.4.2, this issue is examined more closely.

3.4.4.2 Weighted Price Contribution per Trade

The high WPC estimate recorded for the first 2 hours (07:00–09:00 hours) of the RTH coupled with their low level of trading in comparison with the other periods provide the basis to expect high information content per trade during the period. The adverse selection cost component is also highest during this period in the RTH. The study therefore proceeds by examining the information content per trade. As already stated, we use the WPCT measure. The WPC per trading period (interval) is divided by the weighted ratio of trades executed during that period (interval). If for each day, $t_{k,c}$ is the number of executed trades in time period k for contract c, and t_c is the total sum of $t_{k,c}$ for all the periods, then $WPCT_k$ is defined as

$$WPCT_{k,c} = \frac{\left(\frac{|ret_c|}{\sum_{c=1}^{C}|ret_c|}\right) \times \left(\frac{ret_{k,c}}{ret_c}\right)}{\left(\frac{|ret_c|}{\sum_{c=1}^{C}|ret_c|}\right) \times \left(\frac{t_{k,c}}{t_c}\right)}. \quad (3.17)$$

As a consequence of the measure being equivalent to a ratio of the aggregate price shift occurring in a period scaled by the ratio of trades in that same period, the WPCT should be about one if all trades carry similar levels of information to the market. For statistical inference, the standard t-statistic is used to test the null that the daily WPCT values (per period and for each contract) are not significantly different from zero. The Wilcoxon–Mann–Whitney test is also used to obtain statistical inference on differences between the RTH intervals and the AMC period.

The close-to-close WPCT is reported in Table 3.4. The results show that for three of the contracts, the trades in the opening period hold higher levels of information than at any time during the normal trading periods. Some of the estimates for the RTH are noisy as they are not statistically significant. Consistent with Panel A of Tables 3.2 and 3.3, on per contract basis, the contract with the farthest maturity, the December-2012 contract holds the highest level of information per trade. It is also observed that as in Table 3.3, individual trades in the period 15:00–17:00 hours are very informative and are largely statistically significant across all contracts. The informed trading effect of the increasing EFP/EFS trades at this period is more evident as the volume of liquidity-seeking trades starts to taper off. This implies that the level of price discovery reported for this period will have a level of efficiency comparable to that of the AMC period. This is examined in Sect. 3.4.5. The results in this section (Tables 3.3 and 3.4) thus support that the AMC, as well as the opening 2 hours, are very important to the price discovery process.

3.4.5 Efficiency of the Price Discovery Process: Period by Period Analysis

In markets with relatively slim trading volumes like the EU-ETS platforms, big liquidity induced trades are usually associated with short-term price effects that are afterwards reversed. Although the highest proportion of the large trades is in the AMC, results shown so far suggest that there are more liquidity-driven trades in the RTH than in the AMC. Based on this, it is anticipated that the RTH trades will be generally noisier than the AMC trades because of the expected price reversals.

However, since large spreads, as shown in Panel B of Table 3.2 for the AMC period, are typically instrumental to price reversals, there is also the suspicion that there may be an appreciable level of noisy trades in the AMC. Hence the hypothesis here is for lower signal:noise ratio for the RTH than the AMC. Based on foregoing analysis, it is also expected that

Table 3.4 Weighted price contribution per trade by time intervals

	Time periods							
	Normal trading day						AMC	
Contracts	07:00–09:00	09:00–11:00	11:00–13:00	13:00–15:00	15:00–17:00	07:00–17:00	17:00–18:00	Days with zero price change
Dec-2009	−0.12[a]	−0.15	−0.57	−0.68	0.63	−0.12	0.31	0.019
Dec-2010	2.54[a,b]	0.76[a]	−0.72	0.86[a]	0.78[a,b]	0.80[a,b]	−2.90	0.019
Dec-2011	1.96[b]	0.33	1.13[a]	0.83[a]	0.92[b]	0.93[b]	11.54[b]	0.024
Dec-2012	2.78[b]	1.75[b]	1.04[a]	0.63[a]	1.10[b]	1.34[b]	25.53[b]	0.019
Overall	1.79[a]	0.67	0.22	0.41	0.86[a,b]	0.74[a,b]	8.63[b]	0.019

The table shows WPCT of six RTH normal trading day intervals and the AMC period to the close-to-close return for the four highest volume December maturity contracts on the ECX platform. For each contract and interval k the WPCT is computed for each day and then averaged across days:

$$WPCT_{k,c} = \frac{\left(\frac{|ret_c|}{\sum_{c=1}^{C}|ret_c|}\right) \times \left(\frac{ret_{k,c}}{ret_c}\right)}{\left(\frac{|ret_c|}{\sum_{c=1}^{C}|ret_c|}\right) \times \left(\frac{t_{k,c}}{t_c}\right)},$$

$t_{k,c}$ is the number of executed trades in time interval k for contract c, and t_c is the total sum of $t_{k,c}$ for all the intervals. The trading days when close-to-close returns equal 0 are eliminated. The final column shows the fraction of days with close-to-close return equal to 0. Wilcoxon–Mann–Whitney (tie-adjusted) tests are used to determine whether contract-dependent values for normal trading day intervals are significantly different from that of the AMC period

[a] Contract-dependent normal trading day interval during which the contract WPCT is significantly different from that of the AMC
[b] WPCT values significantly different from 0 at the 5% level. The data covers the trading period February 2009 through November 2009. The normal trading day period runs between 7:00:00 and 16:59:59 hours London time; the AMC runs between 17:00 and 18:00 hours London time

the Dec-2011 and Dec-2012 (the most illiquid instruments in the sample) will possess generally low signal to noise ratios (noisy) across all periods. In Sect. 3.4.4.1, the observation that high volatility levels and low WPC estimates reported for the 11:00–13:00 hours period may indicate the presence of noisy trades means that the lowest signal:noise ratio estimates in the RTH is expected for the period. Price efficiency is measured using the so-called unbiasedness regressions; this involves estimating the noisiness of contract prices for different intervals (see Biais et al. 1999).

For each contract and each day, Eq. (3.18) is estimated separately for each time period (60 minutes each for the RTH and 10 minutes each for the AMC), where ret_{cc} is the close-to-close return and ret_{ck} is the return from the close to the end time of period k:

$$ret_{cc} = \alpha + \beta ret_{ck} + \varepsilon k. \tag{3.18}$$

Barclay and Hendershott (2003) argue that the slope coefficient β is a measure of the ratio of price signal to the noise in the pricing process. Considering the regression analysis problem of standard errors-in-variables, assuming contracts returns are accurately computed and they are not correlated, the slope coefficient will be equal to one. Then, take the assumption that the actual return is not observed and also that the observable return is actually equivalent to the real return plus the noise. Noise in this sense refers to microstructure impacts such as spread components or reversible price effects. If one imagines that $ret_{cc} = RET_{cc} + v$ and $ret_{ck} = RET_{ck} + u$, then, consider RET_{cc} and RET_{ck} as the actual returns and u and v have zero mean and respective variances equivalent to u^2 and v^2. An ordinary least squares estimation of Eq. (3.18) will result in the estimated slope coefficient β^*, where

$$\beta^* \xrightarrow{p} \beta \left(\frac{\sigma^2_{RET_{ck}}}{\sigma^2_{RET_{ck}} + \sigma^2_u} \right), \tag{3.19}$$

$\sigma^2_{RET_{ck}}$ is a measure of the total information observed from the previous close to the time period, k and σ^2_u is the noise effect observed in prices at period k. The slope thus measures the ratio of information content (signal) to signal plus noise in prices at period k.

Specifically, a time series estimation of Eq. (3.18) for each contract and each period is conducted. The slope coefficient estimates for each contract is obtained and the mean for all the contracts with respect to each of the time periods calculated. Following Biais et al. (1999), the confidence bands are calculated using the time series' standard errors of the mean of the slope coefficient estimates. As pointed out by Biais et al. (1999), time series estimation of instrument returns in the presence of learning is problematic because of learning-induced non-stationarity. This is relevant to this analysis, especially since it is based in part on learning in the after-hours market. In order to ensure that this analysis does not suffer from the spurious regression problem, individual unit root tests of each time series variable used in the separate regressions are conducted. The test results suggest that the variables are stationary. In any case, the Newey and West (1987) heteroscedasticity and autocorrelation consistent covariance (HAC) matrix estimator, which is consistent in the presence of both heteroscedasticity and autocorrelation of unknown form, is applied. The results obtained from the HAC estimation are materially the same as the ones reported.

Figure 3.3 shows the graph of the signal:noise ratio with the confidence bands. In Table 3.5, there are also two panels of the slope estimates for the RTH and AMC periods. Comparatively, the signal:noise ratio in the RTH is generally lower than the AMC as hypothesised. During the RTH, the signal:noise ratio ranged from about 0.37 to 0.78 and from 0.61 to 0.92 in the AMC, clearly indicating the RTH as being noisier than the AMC period. Ciccotello and Hatheway (2000) and Barclay and Hendershott (2003) find high signal:noise ratios that are sustained over the RTH for the NASDAQ-pre opening in 1996 and 2000, respectively. NASDAQ pre-open in 2000 daily averaged more than $2 million. Biais et al. (1999) instead find low signal:noise ratio for the Paris Bourse that has no official pre-open trading in 1991. Orders are allowed in the pre-open but no execution takes place in 1991 on the Paris Bourse, hence, no volume is registered, although the last 10 minutes before the RTH begins is the most active period for order placement in the day. These facts and the results further underscore the generally held view that trading volumes

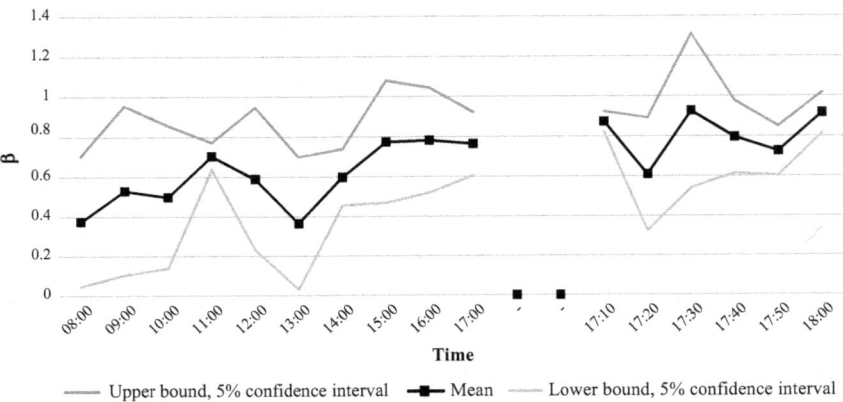

Fig. 3.3 Unbiasedness regressions by intervals. The figure shows the chart of signal:noise ratio over the entire trading periods of RTH and AMC for the December maturity contracts (2009, 2010, 2011 and 2012) trading on the ECX platform. For each contract and each day, the following equation is estimated, using ordinary least squares and Newey and West (1987) HAC, separately for each time period (60 minutes each for the RTH and 10 minutes for the AMC), where ret_{cc} is the close-to-close return and ret_{ck} is the return from the close to the end time of period k. Confidence bands are computed using the time series' standard errors of the slope coefficient estimates.

$$ret_{cc} = \alpha + \beta ret_{ck} + \varepsilon k.$$

The data covers the trading period February 2009 through November 2009. The RTH period runs between 7:00:00 and 16:59:59 hours London time; the AMC runs between 17:00 and 18:00 hours London time

form a vital component of efficient price discovery, especially in thin markets like the EU-ETS platforms. Indeed, the highest trading periods of the day in the sample post the highest signal:noise ratios. This view is strengthened by the fact that a low signal:noise ratio is more likely for the AMC than the RTH because of propinquity to the close.[6] It is therefore proposed that higher trading volumes are associated with higher price efficiency and that this association holds more significance in thin markets.

In addition, our expectation with respect to noisiness of the price discovery process during the 11:00–13:00 hours period is confirmed. The noisiest point of the day according to Fig. 3.3 is at 13:00 hours as a result of the noisiness of price in the less liquid contracts (Dec-2012, Dec-2011 and Dec-2010). The Dec-2012 contract during this period is a very low 0.054, underscoring the fact that a majority of the trades in this contract are very noisy. The estimates suggest that the less liquid a contract is on the ECX platform, the higher the likelihood of its prices being noisy.

The regression estimates for both the RTH and the AMC are obtained using the same close-to-close returns and are therefore correlated. This means the level of statistical difference between the AMC and RTH will be biased using the time series standard errors in Fig. 3.3 and Table 3.5. In drawing statistical inference on the distinction between the slope coefficients, this level of contemporaneous correlation must be considered. We follow Barclay and Hendershott's (2003) method of computing for every day, the difference between the RTH and AMC coefficients and employing the standard error of this time series to draw inferences on the difference between the two periods. The inference is based on the mean difference being significantly different from zero. The result shows that the signal:noise ratio is significantly higher in the AMC than the RTH.

3.5 Chapter Summary

According to Fama (1970), financial market efficiency is a function of the incorporation of the available information to determine instrument price. In this chapter, efficiency of the EU-ETS is inferred through the price discovery process. We analyse the intraday price discovery process on the EU-ETS's largest trading platform and also measure the efficiency of that process. The chapter provides an empirical connection between trading volumes of CFI in the EU-ETS and their contribution to price discovery and informational efficiency. Our analysis shows that the more liquid CFI are, the higher the likelihood that they can be traded efficiently. The price discovery process for the more liquid instruments shows levels of efficiency comparable to those of traditional financial instruments. This is the case during both the normal trading day and the AHT periods in the case of the frequently traded contracts. This is a significant indication of the level of maturity of the EU-ETS. The efficiency of the EU-ETS can therefore provide a basis for the introduction of a global mandatory cap and trade scheme.

The restriction of AMC trading/registration of trades on the ECX to EFP and EFS trades almost entirely ensures that this period on the market

Table 3.5 Unbiasedness regressions by intervals

Contracts		08:00	09:00	10:00	11:00	12:00	13:00	14:00	15:00	16:00	17:00
Panel A: Regular trading hours											
Dec-2009	Estimate	0.85	0.98	0.75	0.66	1.02	0.84	0.71	0.94	0.85	0.95
	Std error	0.099	0.082	0.082	0.082	0.070	0.066	0.052	0.044	0.047	0.024
	t-statistic	8.54[a]	12.01[a]	9.08[a]	8.14[a]	14.56[a]	12.73[a]	13.51[a]	21.39[a]	18.05[a]	38.91[a]
	Adj. R^2	0.26	0.41	0.28	0.24	0.50	0.44	0.47	0.69	0.61	0.88
Dec-2010	Estimate	0.37	0.82	0.83	0.80	0.76	0.36	0.72	0.86	0.91	0.86
	Std error	0.086	0.080	0.061	0.068	0.068	0.067	0.059	0.053	0.031	0.034
	t-statistic	4.31[a]	10.18[a]	13.74[a]	11.79[a]	11.16[a]	5.35[a]	12.33[a]	16.39[a]	29.29[a]	25.02[a]
	Adj. R^2	0.10	0.33	0.47	0.40	0.37	0.12	0.42	0.56	0.80	0.75
Dec-2011	Estimate	0.09	0.20	0.38	0.71	0.37	0.21	0.42	0.98	0.97	0.63
	Std error	0.050	0.063	0.074	0.065	0.065	0.061	0.062	0.044	0.040	0.056
	t-statistic	1.81	3.19[a]	5.12[a]	10.96[a]	5.70[a]	3.35[a]	6.67[a]	22.30[a]	24.22[a]	11.29[a]
	Adj. R^2	0.01	0.04	0.11	0.36	0.13	0.05	0.17	0.70	0.74	0.38
Dec-2012	Estimate	0.20	0.12	0.03	0.65	0.21	0.054	0.55	0.31	0.39	0.63
	Std error	0.043	0.065	0.017	0.067	0.063	0.064	0.057	0.060	0.059	0.059
	t-statistic	4.59[a]	1.81	2.00[a]	9.74[a]	3.40[a]	0.84	9.64[a]	5.23[a]	6.60[a]	10.62[a]
	Adj. R^2	0.09	0.01	0.14	0.31	0.05	0.00	0.31	0.11	0.17	0.35

Contracts		17:10	17:20	17:30	17:40	17:50	18:00
Panel B: After market closes							
Dec-2009	Estimate	0.87	0.98	1.05	0.94	0.91	0.85
	Std error	0.065	0.047	0.043	0.048	0.046	0.054
	t-statistic	13.30[a]	21.04[a]	24.43[a]	19.44[a]	19.62[a]	15.67[a]
	Adj. R^2	0.46	0.68	0.74	0.64	0.65	0.54
Dec-2010	Estimate	0.85	0.67	0.67	0.78	0.67	0.88
	Std error	0.068	0.054	0.076	0.086	0.087	0.0753
	t-statistic	12.53[a]	12.35[a]	8.83[a]	9.01[a]	7.62[a]	11.66[a]
	Adj. R^2	0.68	0.60	0.42	0.45	0.37	0.65
Dec-2011	Estimate	0.95	0.47	0.85	0.93	0.70	1.07
	Std error	0.171	0.083	0.12	0.071	0.172	0.067
	t-statistic	5.56[a]	5.63[a]	7.04[a]	13.10[a]	4.10[a]	15.91[a]
	Adj. R^2	0.48	0.49	0.58	0.83	0.30	0.92
Dec-2012	Estimate	0.83	0.32	0.68	0.54	0.63	0.87
	Std error	0.093	0.082	0.0824	0.084	0.077	0.089
	t-statistic	8.88[a]	3.86[a]	8.28[a]	6.43[a]	8.13[a]	9.71[a]
	Adj. R^2	0.58	0.18	0.51	0.37	0.47	0.68

The table shows results of signal:noise ratio over the entire trading periods of RTH (Panel A) and AMC (Panel B) for the December maturity contracts (2009, 2010, 2011 and 2012) trading on the ECX platform. For each contract and each day, the following equation is estimated, using LS and Newey and West (1987) HAC, separately for each time period (60 minutes each for the RTH and 10 minutes for the AMC), where ret_{cc} is the close-to-close return and ret_{ck} is the return from the close to the end time of period k

$$ret_{cc} = \alpha + \beta ret_{ck} + \varepsilon_k$$

[a] Significance at the 5% level of statistical significance. The data covers the trading period February 2009 through November 2009. The RTH period runs between 7:00:00 and 16:59:59 hours London time; the AMC runs between 17:00 and 18:00 hours London time

will be a session largely for professional traders. Trading EFP/EFS in the AMC period requires the ownership of a platform account. Non-members must retain a clearing member in order to trade at this time. The EFP and EFS trades in the sample are large with more than 81% qualifying as block trades (based on ECX's definition of block trades as trade sizes of 50 lots or more) suggesting that the mechanism is almost an exclusive preserve of institutional investors, compliance participants and other professional traders. Kurov's (2008) analysis of three index futures markets reveals that 61–74% of the price discovery in these markets is from orders initiated by exchange member firms and that this information share exceeds the proportion of trading by them. Therefore, they appear to be the most informed participants in the market. If this class of traders dominates trading in the ECX carbon futures during the AMC market as Table 3.2 suggests, then the results are consistent with the Kurov's (2008) findings (see also Kurov and Lasser 2004). Our conclusion is that there exists a higher level of informed trading in the AMC and that less frequently traded instruments are more likely to have higher levels of informed trading are also consistent with Barclay and Hendershott (2003).

The huge changes in market features observed at the close provide insights into the endogenous impacts of the after-hours market's EFP/EFS trades in the general exchange-traded emissions permit market. The AMC period has the largest volume of contracts traded per minute than the RTH period; however, the reduced total number of trades executed within the period means that each trade becomes very informative. The increase in levels of informed trading starts inside the RTH at about 15:00 hours, accelerates as the closing approaches and eventually peaks during the AMC period. The statistically significant high signal:noise ratios for the contracts in the AMC and the latter part of the normal trading day support this view. A lot of trades during this period are largely executed to take advantage of private information. The restriction placed on the AMC period price movements only serves to make this more remarkable.

Prior to this study, no attention has been given to the implications of EFP and EFS transactions in the EU-ETS. The analysis of these trades in the AMC on the largest carbon exchange in the world provides the first insight into how they alter market characteristics. It also shows the periodic changes in market dynamics during the day and the AMC period.

Perhaps the most fundamental contribution made to the academic literature with this study is the realisation of the unexpected fact that the normal trading day's prices are noisier than the prices discovered outside of RTH, but only on account of the less liquid contracts. That prices are more likely to be reversed in the normal trading day than in the AMC market is yet another confirmation of the high levels of informed EFP/EFS trading activity in the AMC.

The findings here have implications for practitioners and academics alike. For compliance buyers of carbon permits, who must trade in the market or reduce their emissions to avoid regulatory penalties, these results potentially improve confidence in the EU-ETS. Compliance buyers can develop carbon trading strategies with a better understanding of the market price evolution. This includes the distinctions between the different carbon futures instruments and the different trading periods (and intervals).

For investors, this study provides practical insights that can be useful for carbon investment strategies and effective risk management. Investor participation in the market requires the assurance of an appreciable level of price signalling. This is crucial for the efficient allocation of resources, which in turn has welfare effects. By demonstrating that liquid carbon futures enjoy a similar level of informational efficiency to more traditional financial instruments, this study, as presented in this chapter, serves this purpose.

This study also adds to the growing body of literature seeking to understand the carbon trading market in Europe. The possibility of the EU-ETS serving as the platform for establishing a global market led mandatory carbon emissions reduction programme increases the importance of the work being carried out in this area of financial economics research.

Notes

1. Changes, during the day and across days, are in small increments in traditional markets according to Admati and Pfleiderer (1988), Madhavan et al. (1997), Foster and Viswanathan (1993).
2. AMC period on the ECX is restricted to Exchange for Physical (EFP) and Exchange for Swaps (EFS) trades only. A detailed description is provided in Sect. 3.2.

3. The model can be estimated using ML and LS as long as the distributional assumptions are either met or can be accounted for with estimation procedures such as the application of Newey and West (1987) HAC. In this chapter, we opt for LS as is the case with Heflin and Shaw (2000).
4. For the RTH, Eq. (3.14) is estimated using the trade classification provided by ECX in the dataset and also by employing the tick rule (Lee and Ready 1991), estimates obtained from both methods are quantitatively similar. Results based on the tick rule are reported for both RTH and AMC periods.
5. van Bommel (2011) analyse three estimators of price discovery and identified the WPC as consistent, also it is the only unbiased and asymptotically normal measure for price discovery if the price process follows a driftless martingale.
6. Alternate analysis run by Barclay and Hendershott (2003) on the NASDAQ, however, suggests this is unlikely; still it is a possibility that provides additional basis for the argument on the link between trading volume and price discovery efficiency.

References

Admati, A., & Pfleiderer, P. (1988). A Theory of Intraday Patterns: Volume and Price Variability. *The Review of Financial Studies, 1*, 3–40.

Aitken, M., & Frino, A. (1996). The Accuracy of the Tick Test: Evidence from the Australian Stock Exchange. *Journal of Banking & Finance, 20*, 1715–1729.

Almeida, A., Goodhart, C., & Payne, R. (1998). The Effects of Macroeconomic News on High Frequency Exchange Rate Behavior. *The Journal of Financial and Quantitative Analysis, 33*, 383–408.

Andersen, T. G., Bollerslev, T., Diebold, F. X., & Vega, C. (2003). Micro Effects of Macro Announcements: Real-Time Price Discovery in Foreign Exchange. *The American Economic Review, 93*, 38–62.

Barclay, M. J., & Hendershott, T. (2003). Price Discovery and Trading after Hours. *The Review of Financial Studies, 16*, 1041–1073.

Barclay, M. J., & Hendershott, T. (2004). Liquidity Externalities and Adverse Selection: Evidence from Trading after Hours. *The Journal of Finance, 59*, 681–710.

Barclay, M. J., Litzenberger, R. H., & Warner, J. B. (1990). Private Information, Trading Volume, and Stock-Return Variances. *The Review of Financial Studies, 3*, 233–253.

Barclay, M. J., & Warner, J. B. (1993). Stealth Trading and Volatility: Which Trades Move Prices? *Journal of Financial Economics, 34*, 281–305.

Benz, E., & Hengelbrock, J. (2009). *Price discovery and liquidity in the European CO_2 futures market: An intraday analysis*. Working paper presented at the Carbon Markets Workshop, 5 May 2009.

Biais, B., Hillion, P., & Spatt, C. (1999). Price Discovery and Learning During the Preopening Period in the Paris Bourse. *The Journal of Political Economy, 107*, 1218–1248.

van Bommel, J. (2011). Measuring Price Discovery: The Variance Ratio, the R^2 and the Weighted Price Contribution. *Finance Research Letters, 8*, 112–119.

Brock, W. A., & Kleidon, A. W. (1992). Periodic Market Closure and Trading Volume: A Model of Intraday Bids and Asks. *Journal of Economic Dynamics and Control, 16*, 451–489.

Cao, C., Ghysels, E., & Hatheway, F. (2000). Price Discovery without Trading: Evidence from the Nasdaq Preopening. *The Journal of Finance, 55*, 1339–1365.

Cason, T. N., & Gangadharan, L. (2003). Transactions Costs in Tradable Permit Markets: An Experimental Study of Pollution Market Designs. *Journal of Regulatory Economics, 23*, 145–165.

Cason, T. N., & Gangadharan, L. (2011). Price Discovery and Intermediation in Linked Emissions Trading Markets: A Laboratory Study. *Ecological Economics, 70*, 1424–1433.

Chakravarty, S. (2001). Stealth-Trading: Which Traders' Trades Move Stock Prices? *Journal of Financial Economics, 61*, 289–307.

Chan, K., Chung, Y. P., & Johnson, H. (1995). The Intraday Behavior of Bid-Ask Spreads for NYSE Stocks and CBOE Options. *The Journal of Financial and Quantitative Analysis, 30*, 329–346.

Chan, K. C., Christie, W. G., & Schultz, P. H. (1995). Market Structure and the Intraday Pattern of Bid-Ask Spreads for NASDAQ Securities. *The Journal of Business, 68*, 35–60.

Chan, L. K. C., & Lakonishok, J. (1993). Institutional Trades and Intraday Stock Price Behavior. *Journal of Financial Economics, 33*, 173–199.

Choi, J. Y., Salandro, D., & Shastri, K. (1988). On the Estimation of Bid-Ask Spreads: Theory and Evidence. *The Journal of Financial and Quantitative Analysis, 23*, 219–230.

Ciccotello, C. S., & Hatheway, F. M. (2000). Indicating Ahead: Best Execution and the NASDAQ Preopening. *Journal of Financial Intermediation, 9*, 184–212.

Clarke, J., & Shastri, K. (2000). *On Information Asymmetry Metrics*. SSRN eLibrary, Working Paper 251938.

Copeland, T. E., & Galai, D. (1983). Information Effects on the Bid-Ask Spread. *The Journal of Finance, 38*, 1457–1469.

Danielsson, J., & Payne, R. (2010). *Liquidity Determination in an Order Driven Market*. London School of Economics Working Paper, London.

Davies, R. J. (2003). The Toronto Stock Exchange Preopening Session. *Journal of Financial Markets, 6*, 491–516.

Easley, D., Kiefer, N. M., & O'Hara, M. (1996). Cream-Skimming or Profit-Sharing? The Curious Role of Purchased Order Flow. *The Journal of Finance, 51*, 811–833.

Easley, D., Kiefer, N. M., & O'Hara, M. (1997). One Day in the Life of a Very Common Stock. *The Review of Financial Studies, 10*, 805–835.

Easley, D., & O'Hara, M. (1987). Price, Trade Size, and Information in Securities Markets. *Journal of Financial Economics, 19*, 69–90.

Easley, D., & O'Hara, M. (1992). Time and the Process of Security Price Adjustment. *The Journal of Finance, 47*, 577–605.

Fama, E. F. (1970). Efficient Capital Markets: A Review of Theory and Empirical Work. *The Journal of Finance, 25*, 383–417.

Fama, E. F., & MacBeth, J. D. (1973). Risk, Return, and Equilibrium: Empirical Tests. *The Journal of Political Economy, 81*, 607–636.

Flood, M. D., Huisman, R., Koedijk, K. G., & Mahieu, R. J. (1999). Quote Disclosure and Price Discovery in Multiple-Dealer Financial Markets. *The Review of Financial Studies, 12*, 37–59.

Foster, F. D., & Viswanathan, S. (1990). A Theory of the Interday Variations in Volume, Variance, and Trading Costs in Securities Markets. *The Review of Financial Studies, 3*, 593–624.

Foster, F. D., & Viswanathan, S. (1993). Variations in Trading Volume, Return Volatility, and Trading Costs: Evidence on Recent Price Formation Models. *The Journal of Finance, 48*, 187–211.

French, K. R., & Roll, R. (1986). Stock Return Variances: The Arrival of Information and the Reaction of Traders. *Journal of Financial Economics, 17*, 5–26.

Gangadharan, L. (2000). Transaction Costs in Pollution Markets: An Empirical Study. *Land Economics, 76*, 601–614.

Garbade, K. D., & Silber, W. L. (1979). Structural Organization of Secondary Markets: Clearing Frequency, Dealer Activity and Liquidity Risk. *The Journal of Finance, 34*, 577–593.

George, T., Kaul, G., & Nimalendran, M. (1991). Estimation of the Bid-Ask Spread and Its Components: A New Approach. *The Review of Financial Studies, 4*, 623–656.

Glosten, L. R. (1987). Components of the Bid-Ask Spread and the Statistical Properties of Transaction Prices. *The Journal of Finance, 42*, 1293–1307.

Glosten, L. R., & Harris, L. E. (1988). Estimating the Components of the Bid/Ask Spread. *Journal of Financial Economics, 21*, 123–142.

Glosten, L. R., & Milgrom, P. R. (1985). Bid, Ask and Transaction Prices in a Specialist Market with Heterogeneously Informed Traders. *Journal of Financial Economics, 14*, 71–100.

Goodhart, C. A. E., Hall, S. G., Henry, S. G. B., & Pesaran, B. (1993). News Effects in a High-Frequency Model of the Sterling-Dollar Exchange Rate. *Journal of Applied Econometrics, 8*, 1–13.

Greene, J. T., & Watts, S. G. (1996). Price Discovery on the NYSE and the NASDAQ: The Case of Overnight and Daytime News Releases. *Financial Management, 25*, 19–42.

Grüll, G., & Taschini, L. (2011). Cap-and-Trade Properties under Different Hybrid Scheme Designs. *Journal of Environmental Economics and Management, 61*, 107–118.

Gwilym, O., Buckle, M., & Thomas, S. (1997). The Intraday Behaviour of Bid-Ask Spreads, Returns and Volatility for FTSE 100 Stock Index Options. *Journal of Derivatives, 4*, 20–32.

Hansen, L. P. (1982). Large Sample Properties of Generalized Method of Moments Estimators. *Econometrica, 50*, 1029–1054.

Harris, L. (1990). Statistical Properties of the Roll Serial Covariance Bid/Ask Spread Estimator. *The Journal of Finance, 45*, 579–590.

Hasbrouck, J. (1991a). Measuring the Information Content of Stock Trades. *The Journal of Finance, 46*, 179–207.

Hasbrouck, J. (1991b). The Summary Informativeness of Stock Trades: An Econometric Analysis. *The Review of Financial Studies, 4*, 571–595.

He, Y., Lin, H., Wang, J., & Wu, C. (2009). Price Discovery in the Round-the-Clock U.S. Treasury Market. *Journal of Financial Intermediation, 18*, 464–490.

Heflin, F., & Shaw, K. W. (2000). Blockholder Ownership and Market Liquidity. *The Journal of Financial and Quantitative Analysis, 35*, 621–633.

Ho, T., & Stoll, H. R. (1981). Optimal Dealer Pricing under Transactions and Return Uncertainty. *Journal of Financial Economics, 9*, 47–73.

Ho, T. S. Y., & Stoll, H. R. (1983). The Dynamics of Dealer Markets under Competition. *The Journal of Finance, 38*, 1053–1074.

Huang, R. D., & Stoll, H. R. (1994). Market Microstructure and Stock Return Predictions. *The Review of Financial Studies, 7*, 179–213.

Huang, R. D., & Stoll, H. R. (1997). The Components of the Bid-Ask Spread: A General Approach. *The Review of Financial Studies, 10*, 995–1034.

Ibikunle, G., Gregoriou, A., & Pandit, N. (2013). Price Discovery and Trading after Hours: New Evidence from the World's Largest Carbon Exchange. *International Journal of the Economics of Business, 20*, 421–445.

Jiang, C. X., Likitapiwat, T., & McInish, T. H. (2012). Information Content of Earnings Announcements: Evidence from After-Hours Trading. *Journal of Financial and Quantitative Analysis, 47*, 1303–1330.

Joyeux, R., & Milunovich, G. (2010). Testing Market Efficiency in the EU Carbon Futures Market. *Applied Financial Economics, 20*, 803–809.

Kerr, S., & Máre, D. (1998). *Transaction Costs and Tradable Permit Markets: The United States Lead Phasedown.* Motu Economic Research Working Paper, Auckland.

Kim, M., Szakmary, A. C., & Schwarz, T. V. (1999). Trading Costs and Price Discovery across Stock Index Futures and Cash Markets. *Journal of Futures Markets, 19*, 475–498.

Kossoy, A., & Ambrosi, P. (2010). *State and Trends of the Carbon Markets, 2010.* The World Bank Report, Washington, DC.

Kraus, A., & Stoll, H. R. (1972). Price Impacts of Block Trading on the New York Stock Exchange. *The Journal of Finance, 27*, 569–588.

Kurov, A. (2008). Information and Noise in Financial Markets: Evidence from the E-Mini Index Futures. *Journal of Financial Research, 31*, 247–270.

Kurov, A., & Lasser, D. J. (2004). Price Dynamics in the Regular and E-Mini Futures Markets. *The Journal of Financial and Quantitative Analysis, 39*, 365–384.

Kyle, A. S. (1985). Continuous Auctions and Insider Trading. *Econometrica, 53*, 1315–1335.

Lee, C. M., & Ready, M. J. (1991). Inferring Trade Direction from Intraday Data. *The Journal of Finance, 46*, 733–746.

Lin, J., Sanger, G. C., & Booth, G. G. (1995). Trade Size and Components of the Bid-Ask Spread. *The Review of Financial Studies, 8*, 1153–1183.

Linacre, N., Kossoy, A., & Ambrosi, P. (2011). *State and Trends of the Carbon Market 2011.* The World Bank Report, Washington, DC.

Madhavan, A., & Panchapagesan, V. (2000). Price Discovery in Auction Markets: A Look Inside the Black Box. *The Review of Financial Studies, 13*, 627–658.

Madhavan, A., Richardson, M., & Roomans, M. (1997). Why Do Security Prices Change? A Transaction-Level Analysis of NYSE Stocks. *The Review of Financial Studies, 10*, 1035–1064.

Mizrach, B., & Otsubo, Y. (2014). The Market Microstructure of the European Climate Exchange. *Journal of Banking & Finance, 39*, 107–116.

Newey, W. K., & West, K. D. (1987). A Simple, Positive Semi-definite, Heteroskedasticity and Autocorrelation Consistent Covariance Matrix. *Econometrica, 55*, 703–708.

Pascual, R., Escribano, A., & Tapia, M. (2004). Adverse Selection Costs, Trading Activity and Price Discovery in the NYSE: An Empirical Analysis. *Journal of Banking & Finance, 28*, 107–128.

Porter, D. C., & Weaver, D. G. (1998). Post-trade Transparency on Nasdaq's National Market System. *Journal of Financial Economics, 50*, 231–252.

Rittler, D. (2012). Price Discovery and Volatility Spillovers in the European Union Emissions Trading Scheme: A High-Frequency Analysis. *Journal of Banking & Finance, 36*, 774–785.

Roll, R. (1984). A Simple Implicit Measure of the Effective Bid-Ask Spread in an Efficient Market. *The Journal of Finance, 39*, 1127–1139.

Rotfuß, W. (2009). *Intraday Price Formation and Volatility in the European Union Emissions Trading Scheme*. Centre for European Economic Research (ZEW) Working Paper, Manheim.

Stavins, R. N. (1995). Transaction Costs and Tradeable Permits. *Journal of Environmental Economics and Management, 29*, 133–148.

Stoll, H. R. (1978). The Supply of Dealer Services in Securities Markets. *The Journal of Finance, 33*, 1133–1151.

Stoll, H. R. (1989). Inferring the Components of the Bid-Ask Spread: Theory and Empirical Tests. *The Journal of Finance, 44*, 115–134.

Stoll, H. R., & Whaley, R. E. (1990). Stock Market Structure and Volatility. *The Review of Financial Studies, 3*, 37–71.

Uhrig-Homburg, M., & Wagner, M. (2009). Futures Price Dynamics of CO_2 Emission Allowances: An Empirical Analysis of the Trial Period. *Journal of Derivatives, 17*, 73–88.

Van Ness, B. F., Van Ness, R. A., & Warr, R. S. (2001). How Well Do Adverse Selection Components Measure Adverse Selection? *Financial Management, 30*, 77–98.

4

The Price Impact of Block Emissions Permit Trades

4.1 Introduction

Since the EU-ETS was established for the purpose of reducing emissions in industrial installations, a significant proportion of trades on the platforms are institutional (most institutional trades are executed as blocks). The domination of the EU-ETS trading platforms by large traders should be expected given the structure and purpose of the scheme. Trading activity breakdown, however, suggests that if institutional traders do trade on the platform, they do so with stealth, using trade sizes not regarded as block (see discussions in Chap. 3) or they just trade using the OTC mechanisms. OTC trades are the main avenue for institutional trading in the EU-ETS, with one estimate at 70% or more in Euro value (see Rittler 2012).

This study is motivated by the fundamental necessity to understand the impact of the increasingly large number of block trades in the EU-ETS exchange-based futures trading. In 2005, approximately 80% of EU-ETS derivatives trades occurred OTC; most of these trades meet the ECX's definition of block trades. The volume traded OTC progressively decreased

---

The study described in this chapter is based on Ibikunle et al. (2016).

to average approximately 70% of the total transaction value over the entire course of Phase I, the trial period (2005–2007). By January 2010 (during Phase II, the Kyoto commitment period: 2008–2012), the proportion of exchange-based trades in the scheme had reached 50%, according to World Bank estimates (see Kossoy and Ambrosi 2010). This development is driven by the need of participants to avoid counter-party risks, an issue that has taken on greater significance in derivatives markets as a whole.

The price impact of block trades has been extensively researched for equity markets and recently for futures markets (see Chou et al. 2011). Kraus and Stoll (1972) were the first to demonstrate the price impact of block trades. They present several arguments as to the cause of this phenomenon: short-run liquidity effects occurring as a result of price compromise suffered because counter-parties are not readily available; price compromise when instruments are not perfect substitutes for each other, leading to inefficient trading and hence price impact; and the idea that price concessions granted in order to execute market order underscores *desperation* to execute the market order. These factors convey information to markets about the potential value of the order to the counter-parties, and hence the order becomes information driving, leading to price impact. Holthausen et al. (1990) find evidence of premium payment or price concession in the execution of buyer-initiated block trades. They hold that buyers pay a premium prior to a block trade; the premium is consequently incorporated permanently into the price, while no evidence of premium payment is found for block sales. Kraus and Stoll (1972) find that price impact is higher for block purchases than sales, because a concession or implicit commission is usually higher for purchases than sales, suggesting that there is indeed a premium paid on block sales. A major contribution of their pioneering work is the establishment of a positive relationship between block trades and price impact. Chan and Lakonishok (1993), among others (e.g. see Barclay and Warner 1993; Holthausen et al. 1990), provide supporting evidence for this; they also find a relationship between market capitalisation and price impact (see also Chan and Lakonishok 1995).

Holthausen et al. (1987) also investigate price impact due to block trades and discover that larger trades induce larger price impact than smaller trades. Barclay and Warner (1993), Chakravarty (2001) and Alzahrani et al. (2013) also provide evidence that order size and subsequent execution potentially result in corresponding trade price impact. In relation to temporary price impact, the first study to show asymmetry in block trades' price effects is Gemmill's (1996) on the London Stock Exchange, with significant differences in the magnitude of price impact being reported. Gemmill (1996) reports permanent price impact due to block trades on the London Stock Exchange and a permanent impact equivalent to 33% of the bid-ask spread for block trades that are purchased and 17% for block trades that are sold.

Consistent with Gemmill (1996), most of the studies conducted on the price impact of buyer- and seller-initiated block trades report price impact asymmetry between the two groups (see among others Chiyachantana et al. 2004; Chou et al. 2011; Conrad et al. 2001; Holthausen et al. 1990; Keim and Madhavan 1996). They generally submit that prices appreciate after purchase block trades are executed, and depreciate on their sale. The depreciation that occurs after sell trades are executed suffers reversion but purchase block trade induced appreciation remains. Chan and Lakonishok (1993) argue that the reason for this is that block sales have a higher likelihood of involving a broker (acting as an intermediary) than block purchases. The temporary impact from sell trades is therefore a reflection of price concession as compensation for the role played by the broker. Liquidity thus plays a key role in the existence of price impact asymmetry between block purchases and sales (see also Gregoriou 2008).

However, despite the large body of literature on price impact of block trades in the conventional markets, to our knowledge, no study has been undertaken for block trade price impact in permit markets. We therefore attempt an analysis of determinants of price impact in the EU-ETS using tick data from the scheme's largest platform, the ECX, in order to understand the impact of block trades in this important market. Our results show intriguing patterns that are largely inconsistent with earlier studies from traditional markets. For permanent and temporary effects, we find several instances of price impact asymmetry for block purchases and sales. Contrary to previous literature on equity markets, wider spreads lead to smaller price impact. We attribute our findings to the fact that block trades executed after a price run-up induce smaller price impact, as suggested by Saar (2001). The implication is that liquidity concerns in the EU-ETS play a less prominent role in emissions permits pricing than in customary markets. This is supported by the findings in Chap. 3, showing that small amounts of trading lead to larger proportions of price discovery in the EU-ETS. Short-run improvements in liquidity, though an important factor in market efficiency (see Chordia et al. 2008), do not detract from block trade price impact on the world's largest carbon platform. Our findings have implications for compliance traders and policymakers alike. It is important that in designing future phases of the EU-ETS, this and other aspects of our results are considered.

The remainder of this chapter is structured as follows. In Sect. 4.2, we provide a discussion of the EU-ETS mechanism and the setup of the ECX and review related literature based on permit trading. Section 4.3 outlines the data and econometric methodology used. Section 4.4 discusses the empirical analysis, and Sect. 4.5 concludes.

4.2 Background to Study

4.2.1 Institutional Set-up

The ECX is the largest carbon exchange in the world by volume and value, with 92% of EU-ETS on-screen trades registered on the platform in 2010, arguably its most dominant year. The ECX is a member of the Climate Exchange Plc group of companies, which also included the Chicago Climate Exchange (CCX) (until 2010) and the Chicago Climate Futures Exchange (CCFX) (until 2012). It manages the development and marketing of several carbon derivative instruments listed and traded on the Intercontinental Exchange platform (ICE Futures Europe). Investors and compliance buyers trade on the ICE platform, in three different types of derivatives (futures, daily futures and options) that have as underlying either EUAs or CERs.

Trading in the first CFI floated on the platform commenced on Friday, 22 April 2005. This was the ECX EUA Futures contract with December 2005 maturity. Apart from the ECX EUA Futures contracts, there are other CFIs on offer. EUA options contract was introduced on 13 October 2006 and the ECX CER Futures contract on Friday, 14 March 2008 with the December 2008 maturity contract. The options contract variant of the CER units was subsequently introduced on 16 May 2008 More recently (13 March 2009), the exchange introduced EUA and CER Daily Futures contracts, a move interpreted by analysts as an attempt to compete in the carbon spot market which was then dominated by Paris-based Bluenext. Bluenext has since been dissolved. With the exception of the daily futures contracts, all contracts up to June 2013 are listed on quarterly maturity cycles. The 2014–2020 expiration contracts are listed on an annual expiry (December) for the time being. However, trading almost exclusively occurs on the December maturity contracts alone with the nearest December maturity usually accounting for more than 80% of trading volume. A similar observation is made for the EEX study in Chap. 5, while Mizrach and Otsubo (2014) report the same phenomenon for the ECX.

Trading rules and procedures on the exchange follow general industry practice in the more traditional asset classes. Trading commences at 07:00 and continues until 17:00 hours UK local time. There is a pre-trading period of 15 minutes from 06:45 hours to allow members to place orders in preparation for trading start; however, almost no orders are executed during this period. The settlement period, which runs from 16:50:00 to 16:59:59 hours UK time, is the third stage of the trading day and is used for determination of the settlement price. The fourth stage of the trading is the after-hours period reserved only for reporting EFP/EFS trades. In Chap. 3, the contribution of trades reported during this period to price discovery is examined. These trades can be regarded as a form of 'upstairs' trading in the context of the ECX and hence will not be examined in this study since the focus is on downstairs trading only.[1] This follows the precedent of Frino et al. (2007) and Alzahrani et al. (2013).

Trading occurs both directly on the platform and bilaterally off the platform (then registered on the platform for on-screen registration). By virtue of this, the exchange maintains three trading mechanisms: trades occur on the ICE platform ETS, as EFP/EFS transactions, or through the block trade mechanism. A mobile version of the ETS has been made available through a smart phone application called mobile ICE.

Although the larger value of trades occurs off the platform as OTC trades, higher numbers of trades are executed on the platform. Trading on the ICE platform is open only to ICE Futures Europe members with emissions trading privilege and have previously listed a minimum of one trading personnel with the ECX. This person is known to the exchange as a 'Responsible Individual (RI)'. For institutional traders with General and Trade Participant memberships, there is no limit to the number RIs registered with the exchange; however, individual participant members must register only one RI. The RI is the trader known to the exchange and must abide by exchange set rules as well as attain a level of trading competence before admission. Non-members of the exchange can however be involved in order routing by using the ICE platform's front-end application called WebICE and through other similar applications provided by independent software vendors (ISVs) accredited by the exchange. These non-members must be clients of members, who have already gained their consent to use the application for order routing.

Participants submit orders by entering it into the ETS, the trades executed as a consequence of the orders are deemed to be anonymous by exchange rules. The executed trades go via the TRS for allocation to an account. Usually account references are inputted pre-execution; however, this can also be done post-execution. A level of price transparency is ensured by instant posting of real-time prices on ICE platform screens and vendor sources. These vendors include Bloomberg, CQG, E-Signal/FutureSource, Reuters, IDC and ICE Live.

The exchange sets reasonability limits for purchase and sale orders. A purchase (sale) order above (below) the limit is rejected. Sale (purchase) order above (below) the limit is accepted without being executed except the market shifts to alter the reasonability limits and hence places it within the limit. The exchange also maintains a 'no cancellation range' within which trades reported as mistakes may not be cancelled. This rule enhances market confidence and reduces noise trades.

Although trading commences at 7:00 hours UK local time, morning markers are not set until after 09:15 hours; it is the weighted mean of trades occurring within the 15-minute period of 09:00–09:15 hours. And it is calculated for only December maturity EUA Futures contracts. Individual contract indexes are issued after the close of after-hours trading at 18:00 hours for both CER and EUA Futures contracts with December maturities. Open interest, which gives an indication of market depth, is computed at 10:00 hours and released at 11:00 hours UK local time each trading day. It is representative of the previous close's position, adjusted for AHT and transfers/settlements completed before 10:00 hours.

Clearing is provided by Ice Clear Europe, which charges transaction fees on behalf of the exchange. For quarterly contracts, a currently prevailing charge of €0.0035 per tonne of CO_2 (one EUA) for proprietary transactions and €0.004 for non-proprietary transactions are levied on clearing members. The charges are doubled for daily futures contracts. These fees are in addition to annual subscription fees for various participant categories. Transaction fees are not placed on the exercise of an option or on physical delivery of futures contracts. Minimum tick has been held constant at €0.01 per tonne of CO_2 since 27 March 2007 from its previous €0.05 at commencement in 2005.

Trading in contracts cease on the last trading Monday of the expiry month and the underlying are thereafter eligible for delivery within 72 hours of trading cessation. Since Ice Clear Europe acts as the clearing agent, physical delivery of contracts' underlyings are effected through them. By nature of the EU-ETS, clearing members are required to own a person holding account (PHA) with a country registry within the EU-ETS. The transfer of EUAs and CERs are made from the PHA account of the selling clearing member onto the PHA of Ice Clear Europe, then from theirs to the buying party's PHA. All EU registries are currently eligible for physical deliveries under this arrangement. All registries operate on a continuous basis and are connected to the CITL. Transfer of permit rights are done online real time; thus, CITL usually confirms receipt of transfer requests within 60 seconds. However, CITL must conduct further checks to confirm the authenticity of the request and validity of the permits being traded. This can take up to 24 hours. The need for these robust checks has assumed a larger dimension since January of 2011 when more than €30 million worth of permits were stolen from registries across the EU.

The ECX has a defined minimum lot quantity for block trades. Each block trade must be a minimum of 50 lots/contracts (50,000 EUAs or CERs).

4.3 Data and Methodological Approach

4.3.1 Data

Two datasets are obtained for this study; the first is a high-frequency dataset from ICE Futures Europe detailing intraday transactions to the nearest second. The fields contained in this dataset are the time of trade execution to the nearest second, numeric identification for each included CFI, the CFI description, traded month of contract, order identification, trade sign (bid/offer), transaction price, quantity traded and the type of trade (exchange/OTC/Block, etc.). The dataset covers from the start of Phase II EU-ETS (2 January 2008) until the 9 May 2011. The use of the dataset ensures that this study provides the longest time period analysis of Phase II EU-ETS trading of recent years.

The second dataset contains EOD variables, it is also from ICE Futures Europe and covers the same time span and provides daily computations of different variables. The variables are contract specific and include open interest, exchange derived value-weighted indices, settlement price and daily traded volumes.

Only the December expiry contracts are selected because they are the only ones for which official exchange index data is available for; the December maturity contracts are for 2008, 2009, 2010, 2011, 2012 and 2013 expiries. This selection is also based on volume considerations. All trades executed within the initial pre-open period and during the after-hours market are excluded. All other trades executed off-market and in the upstairs market are also excluded. These steps are taken in order to provide a basis for comparing this study's results with previous studies, also in order to ensure robustness and consistency.

After cleaning, the final dataset consists of a total of 961,131 trades over the entire period. This study follows ECX's definition of block trade as any trade with a minimum lot size of 50 contracts (50,000 EUAs). This definition yields a sample size of 16,715 block trades. This is about 1.74% of the total number of trades in the cleaned dataset. The absolute quantity is comparable to the 16,951 NYSE downstairs block trades analysed by Madhavan and Cheng (1997) for 30 Dow Jones stocks and larger than the sample of 5987 from the London Stock Exchange investigated by

Gemmill (1996). Furthermore, considering that permit market trading activities are considerably lower to the traditional asset classes; this can be regarded as a substantial quantity. Also, the percentage of block trades to all other trades is comparable to the Frino et al.'s (2007) adoption of the largest 1% of trades as block trades.

Trade signs allocated to each trade by the exchange are adopted for the final tick data set.[2] Of the 16,715 block trades in the final sample, 8356 are buyer-initiated and 8359 are seller-initiated. As in Frino et al. (2007), the volume of seller-initiated orders is slightly more than that of the buyer-initiated ones.

4.3.2 Methodology

We begin our inquiry by computing three types of price impact generally recognised in the literature. These are the temporary, permanent and total price impact measures. The microstructure literature acknowledges permanent price impacts as those induced by private information and temporary price impacts as those resulting from noise or liquidity-induced trading, leading to reversal of price (see Chan and Lakonishok 1995; Easley et al. 2002; Glosten and Harris 1988).

Usually, block trades demand more liquidity than is likely to be available at current quoted prices. Thus, if a block trade is to be fully executed against the available level of liquidity, it must 'walk' through the order book and as a result ends up forcing instrument prices to shift in the trade direction; that is, down for sells and up for buys. The temporary impact on price measures the market's frictional price reaction to the execution of a block trade, which also dissipates thereafter, hence, the definition represented in Eq. (4.1) below. Specifically, Eq. (4.1) quantifies the liquidity element of price impact since the block trade will be executed at a price different from the equilibrium price as dictated by current quotes. This friction in pricing occurs because of the absence of readily available willing counter-parties that can take the opposite side of the block trade at the best available corresponding quote. The temporary effect can thus be viewed as compensation to the counter-parties providing the liquidity required for block trade execution. The compensation is offered by block purchasers (sellers) as a price premium (discount) in order to entice counter-parties into trading with them.

The permanent impact encapsulates the *enduring* impact of the block trade on an instrument, that is, the price shift that is not reversed following the block trade. Thus, the permanent impact measures the information component of a block trade. This implies that the market has learnt something new about the instrument, which leads to a new price equilibrium. In this study, we follow Holthausen et al. (1990), Gemmill (1996), Frino et al. (2007) and Alzahrani et al. (2013) in using the five-trade benchmark to compute the price impact measures; the equations used are also based on these papers. Thus, for temporary impact (Eq. 4.1), we measure the percentage of price reversal after five trades following the block trade; and for permanent price impact, Eq. (4.2) considers the percentage change in price from five trades prior to the block trade to five trades after the block trade. The third price impact measure, total impact, captures the entire percentage price impact, which includes both the liquidity and the information component. Since, we project (in line with previous studies stated above), that the liquidity effect dissipates only after about five trades, Eq. (4.3) should capture both the temporary and permanent price effects of the block trade. We ensure comparability by calculating all three measures as percentage returns according to Eqs. (4.1), (4.2) and (4.3):

$$ret_{cc} = \alpha + \beta ret_{ck} + \varepsilon_k$$

$$\text{Temporary Impact} = \frac{P_{t+5} - P_t}{P_t} \quad (4.1)$$

$$\text{Permanent Impact} = \frac{P_{t+5} - P_{t-5}}{P_{t-5}} \quad (4.2)$$

$$\text{Total Impact} = \frac{P_t - P_{t-5}}{P_{t-5}} \quad (4.3)$$

We use transaction prices in the absence of direct quotes.[3] We adopt the model of Frino et al. (2007), thereafter employed by Alzahrani et al. (2013), in examining some likely determinants of block trade price impact on the ECX. Accordingly, we estimate the following time series regression with EUA contracts-specific variables:

$$PI_t = \gamma_0 + \gamma_x X_t + \gamma_2 \sum_{i=1}^{9} TD_i + \gamma_3 \sum_{i=1}^{4} DD_i + \gamma_4 \sum_{i=1}^{11} MD_i + \varepsilon_t \quad (4.4)$$

where PI_t corresponds to one of three price impact measures: total price impact, permanent price impact and temporary price impact. The explanatory variables are computed as follows. X_t is a vector of six explanatory variables (size, volatility, turnover, market return, momentum and BAS) defined below. TD_i, DD_i and MD_i are dummy variables for time (hour) of day, day of week and month of year and are further defined below.

Size represents the natural logarithm of volume of contracts contained in the block transaction.[4] Based on the assumption that trade size corresponds to information content (see Chan and Lakonishok 1993, among others; Easley and O'Hara 1987; Kraus and Stoll 1972), we adopt trade size as a proxy for information content of the block trade. When investors have private information about an instrument, they act based on the new belief developed as a result of the new information. Hence, they place a sell order if the belief is that the instrument is overpriced, or purchase if the instrument is underpriced (see also Madhavan et al. 1997). *Volatility* represents the standard deviation of trade execution price returns for the trading day up until the block trade.[5] This measure is in line with previous studies (e.g. see Frino et al. 2007). Volatility is representative of intraday fluctuation in trading prices; it shows the pattern of trading belief over the course of a trading session and can therefore be regarded as an implicit proxy of adverse selection costs of trading. The higher the level of volatility of an instrument, the greater the risk associated with it, thus leading to wider spreads as compensation for trading (see Sarr and Lybek 2002). The onset of larger spreads on account of volatility suggests that volatility will lead to price impact. It is expected that volatility of the futures contracts will be positively related to price impact (Domowitz et al. 2001).

Turnover represents the natural logarithm of the aggregate Euro value of all futures contracts traded on the trading day prior to the execution of the block trade, divided by the prevailing Euro volume of open interest.

Turnover has been regularly employed as a measure of trading activity and market liquidity (see among others Frino et al. 2007; Hu 1997; Lakonishok and Lev 1987). Further, open interest has been established as a component of market liquidity measures in futures markets. Using open interest as a component of the proxy for market depth (liquidity) follows Bessembinder and Seguin (1992) and Fung and Patterson (1999). Open interest is a reflection of the order flow of trades and the readiness of traders to risk their funds, and therefore has similar levels of correlation with volatility that spreads have. Price impact is expected to be lower with improvements in liquidity; hence, we anticipate a negative relationship with price impact.[6] *Momentum* is computed as the lagged cumulative daily return for each contract over five trading days before the trading day of the block trade. This expresses the trading trend for the specific instrument. Higher returns will indicate a purchasing trend, and lower returns, a selling trend. Saar (2001) argues that the price performance history of an instrument is related to its expected price impact asymmetry. Specifically, block trades that are executed on the back of decreasing price performance will manifest higher positive asymmetry and block trades executed after a strong run of price appreciation should exhibit less impact or possibly negative asymmetry. Since the transition to Phase II in the EU-ETS, the market has experienced stronger liquidity and market efficiency, and hence has largely been on a run-up in terms of price performance. Based on this, we anticipate momentum will have predominantly negative price impact coefficients.

BAS is a second measure of liquidity in the model. Relative bid-ask spread is the prevailing relative bid-ask spread when the block transaction is executed. We expect that when spreads are wide, there would be higher price impact than when they are narrow. We compute relative bid-ask spread as the last ask price prior to the block trade minus the last bid price before the block trade, divided by the midpoint of both prices. *Market return* is the contract-specific daily return on the ECX EUA index for each contract. By adopting contract-specific return we emulate Frino et al. (2007) in using a more refined measure of market return. Alzahrani et al. (2013) and Frino et al. (2007) report intraday effects for block trade impact. Thus, for consistency, we introduce trading hour, day of week and month of year dummy variables in order to capture trading time/period effects. For these sets of dummies, the last trading hour (16:01–17:00), Friday and December are employed as references. The use of December as a reference month is important given that the contracts in our sample all have December expiries.

4.4 Results and Discussion

4.4.1 Descriptive Statistics

Panel A in Table 4.1 shows descriptive statistics based on trade classification. Of the 16,715 block trades in our final sample, 8356 are buyer-initiated and 8359 are seller-initiated. The total volume of block trades to the total number of trades in the sample is 1.74%. In comparison to conventional markets, trading in a permit market like the ECX seems to be less dependent on institutional activity. However, this is only true if we equate block trading activity to institutional activity. The nature of the EU-ETS is such that emissions are capped and traded in the upstream; hence, trading in EU-ETS permits is dominated primarily by both installations trading for compliance purposes and other institutional investors, such as Barclays Capital. However, most institutional trading occurs OTC and in the upstairs market. While comprising only a small portion of EUA trades on the ECX, institutional trades account for a far higher proportion of the Euro volume of the exchange-based trades (see Mizrach and Otsubo 2014 and findings in Chap. 3). The 0.869% of all the trades in the final sample are identified as buyer-initiated block trades, while a marginally higher percentage of 0.87% are seller-initiated block trades. This trend, while conforming to some previous studies (e.g. Frino et al. 2007), contrasts with others (e.g. see Gregoriou 2008).

After removing the high-volume EFP/EFS trades from the on-screen block trades, we have a total of 16,715 block trades with a combined value of approximately €21 billion. For all block trades, the average number of contracts per trade is more than 13 times the value of all the trades combined (both block and non-block). The average number of trades (transaction value) for block purchases is higher than sales, at 80.21 (€1,258,030) and 77.67 (€1,207,400), respectively. The average relative bid-ask spread

Table 4.1 Summary statistics for block trades

Panel A: Summary statistics for block trades						
	Number of trades	Average number of contracts/trade	% of total number of trades	Average transaction value/trade (€'000)	Average relative spread (%)	Standard deviation (%)
All trades	961,131	6.79		108.40	0.07	
Block trades	16,715	78.94	1.74	1232.71	0.06	0.47
Block purchases	8356	80.21	0.87	1258.03	0.07	0.58
Block sales	8359	77.67	0.87	1207.40	0.06	0.30

Panel B: Correlation matrix for determinants						
	BAS	Market return	Momentum	Volatility	Size	Turnover
BAS	1					
Market return	−0.017	1				
Momentum	−0.023	0.261	1			
Volatility	0.322	0.001	−0.041	1		
Size	−0.012	0.016	−0.002	−0.019	1	
Turnover	−0.080	0.076	0.152	−0.257	0.006	1

Panel A shows descriptive statistics for block trades of December maturity EUA futures executed on the ECX platform between January 2008 and April 2012. Panel B shows the correlation matrix of the determinants of price impact of block trades, the determinants/variables are as defined in Table 4.2

value is 0.067% for purchases and 0.056% for sales. The average relative bid-ask spreads for *all trades* compare favourably with those of all block trades. With the exception of block purchases, the spread for all trades is higher than all classes in Table 4.1. For more developed markets and traditional asset classes, the expectation would be to have reduced spreads for all trades and larger spreads for block trades, since they are more likely to be influenced by private information rather than the search for liquidity. A number of microstructure studies suggest that large-sized trades are more informative than smaller ones (see Easley and O'Hara 1987 for further details). Investors have been known to fragment trades over a period of time in order to take advantage of private information rather than execute an *abnormally* large trade; they do this in order to avoid revealing the privately held information before they can take advantage of it. To some extent, the estimates in Table 4.1 showing block purchases with a higher average number of contracts per trade seem to confirm this intuition. It is also noted that *all trades*, which is approximately 11 times smaller than the average block trade, has a slightly higher average relative spread. A possible explanation is the noisy nature of price discovery during the trading day on the ECX. Noise in the price discovery process and information asymmetry on the ECX have been documented in Chap. 3. In any case, the statistics presented in Table 4.1 are not by themselves conclusive evidence of noise in our data. Also, if noise is synonymous with trading on a platform, we must ensure that the data are representative of this fact; results presented in Sect. 4.5 will provide clearer insights.

4.4.2 Regression Results and Discussion

4.4.2.1 Price Impact and Trade Sign

We investigate the determinants of price impact of block trades for all, purchase and sell transactions separately in this section; Table 4.2 shows the results. Size coefficient estimates for sell block trades are all highly statistically significant and negative. The coefficients confirm that larger sell block trades have both permanent and temporary impacts on the price of carbon futures on the ECX, implying that the larger the block trade, the bigger the negative sell block trade impact. The temporary effects, however, contradict expectations, since price should fall after a sell trade; further analysis examines this curious relationship. There are further and substantial instances of this relationship evident in the results given in Table 4.2. On the one hand, there is sufficient evidence to show that block trades on the ECX induce statistically significant price impact; on the other hand, these impacts do not conform to the established literature stream for conventional markets. For example, the total effects of block purchases are negative and statistically significant for all of volatility, market return, momentum and BAS. The corresponding total effects coefficients for combined block trades are also negative and statistically significant for all of those variables, with the exception of volatility. This implies that purchase block trades dominate the direction of block trade impact, since the sell

Table 4.2 Determinants of price impact of block trades

Variables	Permanent effects			Total effects			Temporary effects		
	All	Purchase	Sell	All	Purchase	Sell	All	Purchase	Sell
Size	−3.25E-05	−1.47E-05	−0.0004***	−1.78E-05	0.0004***	−0.0002**	−1.34E-05	−2.68E-05	0.0001***
	(1.74E-05)	(2.59E-05)	(0.0002)	(1.04E-05)	(0.0001)	(9.84E-05)	(1.73E-05)	(2.65E-05)	(1.17E-05)
Volatility	0.1820*	0.0817	0.0399**	0.0317	−0.0417**	0.0285***	0.0735**	0.1396***	−0.0109
	(0.0944)	(0.1318)	(0.0188)	(0.1201)	(0.0200)	(0.0080)	(0.0377)	(0.0391)	(0.0673)
Turnover	−0.0733	−0.0740	−0.1817***	−0.1357*	−0.2657**	0.0340	0.0530***	0.1730*	−0.1158***
	(0.1367)	(0.2075)	(0.0552)	(0.0801)	(0.1211)	(0.0946)	(0.0129)	(0.0944)	(0.0193)
Market return	0.0157***	0.0102	0.0174***	−0.0147***	−0.0248***	0.0124***	0.0204***	0.0350***	0.0150***
	(0.0060)	(0.0099)	(0.0054)	(0.0046)	(0.0080)	(0.0032)	(0.0070)	(0.0124)	(0.0049)
Momentum	−0.0031	−0.0041*	−0.0020	−0.0028**	−0.0040***	−0.0016**	−0.0005	−0.0003	−0.0004
	(0.0020)	(0.0025)	(0.0023)	(0.0013)	(0.0015)	(0.0013)	(0.0025)	(0.0040)	(0.0019)
BAS	−0.08107	−0.0833	0.0649**	−0.1994**	−0.1006***	0.1318***	0.0273	0.1291	0.0319
	(0.0641)	(0.0544)	(0.0306)	(0.0924)	(0.0415)	(0.0370)	(0.0946)	(0.1266)	(0.0770)
TD_1	0.0004	0.0004	0.0003	−9.53E-06	−0.0006*	0.0003*	0.0004	0.0009	5.45E-05
	(0.0005)	(0.0009)	(0.0003)	(0.0002)	(0.0003)	(0.0002)	(0.0005)	(0.0010)	(0.0002)
TD_2	0.0006	0.0012	−2.74E-05	−0.0002	−0.0002	−8.88E-05	0.0008	0.0014	6.17E-05
	(0.0005)	(0.0010)	(0.0002)	(0.0002)	(0.0002)	(0.0001)	(0.0005)	(0.0010)	(0.0002)
TD_3	0.0006	0.0010	0.0002	−7.30E-05	−0.0003	9.91E-05	0.0007	0.0012	0.0001
	(0.0005)	(0.0009)	(0.0003)	(0.0002)	(0.0003)	(0.0001)	(0.0005)	(0.0009)	(0.0002)
TD_4	0.0006	0.0011	5.40E-05	−0.0002	−0.0005*	0.0001	0.0008	0.0016*	−5.84E-05
	(0.0005)	(0.0009)	(0.0002)	(0.0002)	(0.0003)	(0.0001)	(0.0005)	(0.0009)	(0.0002)
TD_5	0.0002	0.0008	−0.0005**	−0.0005***	−0.0008***	−0.0003**	0.0006	0.0016*	−0.0002
	(0.0005)	(0.0009)	(0.0003)	(0.0002)	(0.0003)	(0.0001)	(0.0005)	(0.0009)	(0.0002)
TD_6	0.0008*	0.0013	0.0004	0.0003	0.0003	0.0003*	0.0005	0.0010	9.21E-05
	(0.0005)	(0.0009)	(0.0003)	(0.0003)	(0.0005)	(0.0002)	(0.0005)	(0.0010)	(0.0002)
TD_7	0.0007	0.0012	0.0001	−0.0002	−0.0005	7.00E-05	0.0009*	0.0017	7.43E-05
	(0.0005)	(0.0009)	(0.0002)	(0.0002)	(0.0003)	(0.0001)	(0.0005)	(0.0009)	(0.0002)
TD_8	0.0002	0.0007	−0.0005*	−0.0003**	−0.0004*	−0.0003**	0.0005	0.0011	−0.0002
	(0.0005)	(0.0009)	(0.0002)	(0.0001)	(0.0002)	(0.0001)	(0.0005)	(0.0009)	(0.0002)

TD_9	0.0005	0.0010	−0.0001	−0.0003**	−0.0004**	−0.0002	0.0007	0.0013	8.75E-05
	(0.0005)	(0.0009)	(0.0002)	(0.0001)	(0.0002)	(0.0001)	(0.0005)	(0.0009)	(0.0002)
DD_1	5.48E-05	−0.0002	0.00023	−0.0002	−0.0004*	3.21E-05	0.0002	0.0003	0.0002
	(0.0005)	(0.0010)	(0.0003)	(0.0002)	(0.0003)	(0.0001)	(0.0005)	(0.0010)	(0.0002)
DD_2	0.0003	0.0005	2.96E-05	−0.0002*	−0.0004	−0.0001	0.0005	0.0009	0.0002
	(0.0004)	(0.0008)	(0.0002)	(0.0001)	(0.0002)	(0.0001)	(0.0004)	(0.0008)	(0.0002)
DD_3	0.0004	0.0007	7.64E-05	0.0001	7.70E-05	0.0001	0.0003	0.0007	−4.50E-05
	(0.0004)	(0.0008)	(0.0002)	(0.0002)	(0.0002)	(0.0001)	(0.0004)	(0.0008)	(0.0002)
DD_4	0.0001	0.0004	−0.0003	−0.0002*	−0.0002**	−4.31E-05	0.0003	0.0009	−0.0002
	(0.0004)	(0.0008)	(0.0002)	(0.0001)	(0.0002)	(0.0001)	(0.0004)	(0.0008)	(0.0002)
MD_1	0.0005	0.0003	0.0007**	0.0005***	0.0003	0.0008***	−1.76E-05	5.21E-05	−9.67E-05
	(0.0003)	(0.0006)	(0.0003)	(0.0002)	(0.0003)	(0.0002)	(0.0003)	(0.0005)	(0.0003)
MD_2	0.0007**	0.0005	0.0009***	0.0006***	0.0007	0.0006***	6.07E-05	−0.0002	0.0003
	(0.0003)	(0.0005)	(0.0003)	(0.0003)	(0.0005)	(0.0002)	(0.0003)	(0.0005)	(0.0003)
MD_3	0.0005*	0.0002	0.0008**	0.0002	−7.33E-05	0.0005**	0.0003	0.0002	0.0003
	(0.0003)	(0.0005)	(0.0003)	(0.0002)	(0.0003)	(0.0002)	(0.0003)	(0.0005)	(0.0003)
MD_4	−8.85E-06	−0.0006	0.0007**	0.0003*	8.71E-05	0.0006***	−0.0003	−0.0007	8.85E-05
	(0.0007)	(0.0013)	(0.0003)	(0.0002)	(0.0003)	(0.0002)	(0.0007)	(0.0013)	(0.0003)
MD_5	−8.92E-05	−0.0005	0.0004	0.0002	−8.97E-05	0.0005**	−0.0003	−0.0004	−0.0001
	(0.0003)	(0.0005)	(0.0003)	(0.0002)	(0.0003)	(0.0002)	(0.0003)	(0.0005)	(0.0003)
MD_6	0.0007**	0.0005	0.0010***	0.0006***	0.0004	0.0008***	0.0002	0.0001	0.0002
	(0.0003)	(0.0005)	(0.0004)	(0.0002)	(0.0003)	(0.0002)	(0.0003)	(0.0005)	(0.0003)
MD_7	0.0007*	0.0001	0.0013***	0.0003	−0.0001	0.0008***	0.0004	0.0003	0.0005*
	(0.0004)	(0.0013)	(0.0004)	(0.0002)	(0.0004)	(0.0002)	(0.0003)	(0.0013)	(0.0003)
MD_8	0.0005	0.0006	0.0005	0.0004**	0.0003	0.0006***	0.0001	0.0003	−0.0001
	(0.0004)	(0.0005)	(0.0004)	(0.0002)	(0.0003)	(0.0002)	(0.0003)	(0.0005)	(0.0004)
MD_9	0.0004	0.0004	0.0005*	0.0004*	0.0003	0.0006***	2.49E-05	0.0002	−8.87E-05
	(0.0003)	(0.0005)	(0.0003)	(0.0002)	(0.0005)	(0.0002)	(0.0003)	(0.0005)	(0.0003)
MD_{10}	0.0004	0.0002	0.0005	0.0002	−6.68E-06	0.0004*	0.0002	0.0002	7.57E-05
	(0.0004)	(0.0005)	(0.0004)	(0.0002)	(0.0003)	(0.0002)	(0.0003)	(0.0005)	(0.0003)

(continued)

Table 4.2 (continued)

MD_{11}	-0.0004	-0.0011	0.0006*	0.0003	0.0002	0.0005**	-0.0007	-0.0013	8.02E-05
	(0.0007)	(0.0013)	(0.0003)	(0.0002)	(0.0003)	(0.0002)	(0.0007)	(0.0005)	(0.0003)
Constant	-0.001	-0.0018	-0.0002	-0.0001	-1.25E-05	-0.0002	-0.0010	-0.0019	4.32E-05
	(0.0007)	(0.0013)	(0.0004)	(0.0003)	(0.0005)	(0.0002)	(0.0007)	(0.0014)	(0.0003)
Observations	16,715	8356	8359	16,715	8356	8359	16,715	8356	8359
R^2	0.0022	0.0024	0.0230	0.0072	0.0218	0.0269	0.0026	0.0044	0.0081

The table reports regression results for all purchase and sell block trades of December maturity EUA futures contracts executed on the ECX platform between January 2008 and April 2011. The coefficients are reported along with the standard errors (in parenthesis). The following regression is estimated using OLS with Newey and West (1987) heteroscedastic and autocorrelation consistent covariance matrix:

$$PI_t = \gamma_0 + \gamma_x X_t + \gamma_2 \sum_{i=1}^{9} TD_i + \gamma_3 \sum_{i=1}^{4} DD_i + \gamma_4 \sum_{i=1}^{11} MD_i + \varepsilon_t$$

where PI_t corresponds to one of three price impact measures: total price impact, permanent price impact and temporary price impact. The explanatory variables are computed as follows. X_t is a vector of six explanatory variables (size, volatility, turnover, market return, momentum and BAS) defined below. TD_i, DD_i and MD_i are dummy variables for time (hour) of day, day of week and month of year and are further defined below. Size represents the natural logarithm of the number of December maturity futures contracts for each block trade; volatility is the standard deviation of trade-to-trade returns prior to the block trade on the trading day; Turnover is the natural logarithm of the futures contracts turnover on the trading day prior to the block trade, turnover is the ratio of total trade volume prior to the block to the prevailing open interest estimates; market return is the return of EUA Futures contract specific index computed by the ECX; momentum corresponds to the cumulative return on the specific EUA Futures contract in the five days prior to the block trade; BAS is the prevailing relative bid-ask spread at the time the block trade is executed, BAS is measured as the last ask price prior to the block trade minus the last bid price before the block trade, divided by the average of both prices. TD_1 to TD_9 equal 1 if the block trade occurs in any of the corresponding hour of trade from the first hour (1) to the ninth hour (9) of the trading day and 0 otherwise. DD_1 to DD_4 equal 1 if the block trade occurs in any of the corresponding day of the week from Monday (1) to Thursday (4) of the trading week and 0 otherwise. Any of MD_1 to MD_{11} equals 1 if the block trade occurs in any of the corresponding months of the year from January (1) to November (11) and 0 otherwise. One EUA Futures contract has an underlying of 1000 EUAs

***, ** and * indicate statistical significance at 1, 5 and 10% level, respectively

block trade total impact coefficients are mostly positive. However, only a fraction of the values obtained support this conclusion. It is misleading to focus on the total price impact estimates, since it is difficult to tell which of the two key impact types (liquidity/temporary and information/permanent) is predominant. For clarity, we examine the temporary and permanent impact estimates in Table 4.2. Here, the block purchase price effects are more in keeping with existing literature. For example, the positive (and statistically significant) temporary effects volatility and market return estimates for the purchase trades are consistent with theory and the current literature (e.g. see Alzahrani et al. 2013; Chiyachantana et al. 2004).[7] The literature suggests that increased depth (turnover) reduces block trade price impact. Our results imply that this is only supported for purchases in the case of total price impact. Temporary effects estimates for both purchases and sales suggest that for liquidity-induced trades, market depth does not dull trade impact: the direction of impact remains consistent. Only the block sale coefficient is significant for permanent effects, contradicting Frino et al. (2007), but supporting Alzahrani, Gregoriou and Hudson's (2013) work on the Saudi Stock Market (SSM). Alzahrani, Gregoriou and Hudson (2013) suggest that huge block sell trades in actively traded instruments may signal adverse information about the intent of the trades, since they indicate beliefs of informed participants and consequently lead to an increase in instrument sales. This can lead to an intensification of the price impact of the block trades involved.

Positive market return coefficient estimates indicate larger price impact for purchase block trades and reduced impact for block sell trades; results obtained are largely in keeping with this expectation. Consistent with Frino, Jarnecic and Lepone (2007) and Alzahrani, Gregoriou and Hudson (2013), they are positive for the full range of price effects for the block sell trades. The permanent effects and temporary effects estimates for the purchase trades are positive as well. According to Chiyachantana et al. (2004), institutional block trades executed on the back of price appreciation lead to lesser permanent price impact. This corroborates Saar (2001), which reports that block trades executed following a recent price run-up generate smaller price impact. The total effects momentum coefficient estimates show statistically significant results. The total effects coefficient estimate for purchase block trades is negative and significant, indicating lesser price impact as a result of price run-up. The negative and statistically significant sell coefficient (total effects) implies the opposite trend holds for sell block trades. The statistically significant BAS estimates (purchase: total effects; sell: total and permanent effects) imply that with wider spreads, there is reduced price impact for both purchase and sell trades, and thus with narrower spreads, there is greater price impact. This contradicts previous studies (e.g. see Aitken and Frino 1996; Gemmill 1996). The ECX is a platform created for trading emission permits, unlike equity markets aimed at trading stocks. Emission permits are designed to be submitted only once a year, but the market remains very liquid all year round since the commencement of Phase II of the EU-ETS (see Montagnoli and de Vries 2010 and Chap. 3). Trading emission permits when they are not immediately needed for submission indicates a level of informed trading. Based on this reasoning, high levels of liquidity may not necessarily inhibit price impact for block trades. The BAS estimates support those obtained for the Turnover variable.

Buyer- and seller-initiated block trades on the ECX induce both temporary and permanent price impacts. Increased liquidity also generates a larger price impact. The dissimilarities in market properties between the EU-ETS cap and trade scheme and traditional financial markets is underscored by the differences in the impact of the two liquidity measures used in the model. Evidence also confirms the prediction that there is a reduced price impact for both purchase and sell block trades when an instrument is on a price run-up prior to the block trade execution.

4.4.2.2 Intraday Variations in Price Impact

In conventional markets, it has been reported that spreads conform to a U-shaped pattern over the trading day. However, little has been reported on intraday variations in the EU-ETS. Rotfuß (2009) reports a roughly U-shaped pattern of intraday volatility on the ECX using ECX data from the first year of trading in Phase II (2008). In Chap. 3, we report a slightly different inverted S-shaped intraday pattern of volatility and trading volume using data from the same platform. To our knowledge, the only available evidence of intraday variations in estimated spread pattern available for the EU-ETS is based on the sample we employ for the analysis in

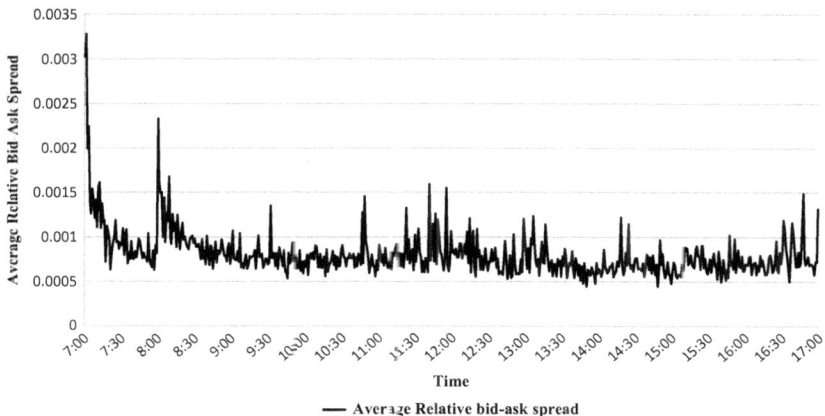

Fig. 4.1 Intraday variations in relative bid-ask spread on the ECX. The figure shows intraday relative bid-ask spread pattern for all trades of December maturity EUA Futures contracts executed on the ECX platform between January 2008 and April 2011. Average bid-ask spread which is defined for each trade as the last ask price prior to the trade minus the last bid price before the trade divided by the average of both prices is computed for each of the six EUA Futures contracts in our sample and then averaged cross-sectionally across all contracts

Chap. 3. Figure 4.1 further shows intraday variations in the average relative bid-ask spread, computed by using our entire dataset (including non-block trades). There is a discernible suggestion of a U-shaped pattern emerging.

In order to examine the presence of intraday variations in the intensity of block trade price impact for the ECX, we introduce intraday dummies for each trading hour, and the results are also presented in Table 4.2.[8] The intervals are as defined in Sect. 4.4, and Table 4.2 also contains the relevant results. Results suggest that block trades executed during some intervals of the trading day generate price impacts, which are significantly different from those, generated by trades executed during the last hour of the day. The middle of the trading day, between 11 am and 3 pm, shows a propensity for inducing a larger sell block price impact than the last hour of the trading day. The difference in the effect of intraday trading activity patterns on the ECX and other traditional financial platforms is underscored by this behaviour. Some studies show that the first hour of trading has been reported as the period when block trades induce the largest price impact (see Frino et al. 2007 as an example). Earlier in this section, we report that, contrary to earlier studies, wider spreads in fact characterise less price impact on the ECX. This result is therefore the only logical outcome of our investigation of time of day effect. This is because Fig. 4.1 shows that the lowest spreads are experienced during the middle of the day, and the highest spread intervals happen during the first hour of the trading day. There is, however, some evidence of price impact asymmetry, since some of the block purchase coefficients are negative and significant. This trend suggests that during the 4th, 5th, 8th and 9th hours of the trading day, block purchases induce less total price impact than for the last hour of the trading day. Further examination of the results shows that for temporary effects, which proxies the liquidity component of price shifts, there is no evidence of the price impact asymmetry observed for total effects. This set of results underscores the observation in Chap. 3 that the last hour of trading on the ECX is dominated by informed trading and thus is likely to exhibit larger levels of permanent price impact due to new information being impounded into the prices. This is because the final trading hour is largely dominated by purchase traders; this also explains the price impact asymmetry with block sells.

4.4.2.3 Day of the Week Effects

We compute and compare mean price impact measures for each day of the week and each month of the year. We observe weak but significant differences in price impact in the day of the week analysis. This leads us to expect some level of price impact variations on account of trading day; hence, the decision to include day of the week dummies in Eq. (4.4) The month of the year analysis yield stronger evidence of significant differences, therefore, we also include month dummies in Eq. (4.4); December, the delivery month for all the futures used in this study, is employed as the reference month. Results presented in Table 4.2 suggest that while there are some statistically significant differences in the price impact of block trades on account of trading day of the week, these are restricted to block purchases only and only exist in the case of total effects. Further, the level of statistical significance is generally weak. Thus, we document another instance of price impact asymmetry on the basis of block trade type.

Table 4.2 also presents the results for the month of year dummies. The results are quite interesting: we observe strong price impact asymmetry between block purchases and sells. For total effects, all 11 sell block month dummies are statistically significant and eight are for permanent effects. None of the purchase block month dummies meet the conventional level of statistical significance. Cases of observed price impact asymmetry between trade types are quite common in microstructure literature; in the conclusion to this chapter, we attempt an explanation of this phenomenon as it applies to this unusual market. The results obtained for block sells strongly suggest that the price impact for sell block trades is highest in December. Since we eliminate December trades for maturing contracts from our sample, this is expected. The elimination of the December trades of the contract closest to maturity means that December trades may be relatively less volatile than trades in the other months of the year. The volatility coefficients in Table 4.2 suggest that there is reduced price impact with increasing volatility in the market. Thus, other months, being relatively more volatile than December, would experience less sell block trade impacts.

4.4.2.4 Trade Size Dependencies on Price Impact

Microstructure studies suggest that liquidity influenced trades are usually characterised by small orders while informed trades usually have larger orders (see Glosten and Harris 1988 as an example). Analyses of the large orders dominated AHT market on the ECX in Chap. 3 indicate that this proposition may be valid in the European carbon market. If different types of trades are characterised by differing trades sizes, it is suspected that block trades will not uniformly cause price impact. Price impact as an important function of trade size has already been demonstrated in this book. In order to determine how block trades of different sizes can affect price functioning, the approach of Alzahrani et al. (2013) and Madhavan and Cheng (1997) in dividing block trades in the sample into three different trade size categories is adopted. Block trades are divided into three groups as follows: 50,000–100,000 EUAs (small), 100,001–200,000 EUAs (mid) and >200,000 EUAs (large). Equation (4.4) is estimated for each of the groups using all three measures of price impact already identified.

Table 4.3 reports the results for purchase block trades. The first observation we make is that of a very high proportion of small block trades. Approximately, 90% of executed trades in the sample are for small blocks. Another observation is the dearth of estimates significantly different from zero. In addition, the results are largely consistent with Table 4.2, especially for the small blocks, since they dominate the sample. However, a few observations deserve mention. As stated earlier, positive market return induces greater permanent price impact, there is now further evidence that this is linked to trade size. The large purchase block trades on average effect 26.26% more permanent price shifts when market return is positive than the small purchase block trades; this is because larger trades convey more information to the market. Conversely and expectedly, owing to the fact that temporary price impacts are composed of liquidity effects, the small purchase block trades cause on average 78.28% more temporary price shifts than the large trades. It is also noted that the large coefficient estimate for temporary effects is not even significantly different from zero.

We also observe that the larger blocks, especially the mid-size blocks, potentially induce higher price impact, in line with theory, than smaller-sized ones. For example consider the volatility variable, which measures the dispersion of participants' belief, its (volatility) total effects estimates for all three trade sizes are statistically significant. The coefficient of the mid-size blocks is positive and statistically significant, and also higher than the other groups at 1.44. This means that for this group, increasing volatility leads to higher price impact. The negative and statistically significant coefficients for the small and large blocks imply that increased volatility does not necessarily imply higher price impact; it instead signals the opposite. This is consistent with earlier total effects coefficient estimates for purchase block trades from Table 4.2. Since these two groups account for more than 93% of the sample size, the consistency with earlier results on purchase trades is not surprising, but the mid-size blocks estimate is consistent with theory. The total effects coefficient estimates

Table 4.3 Determinants of price impact and block trade sizes (purchases)

	Permanent effects			Total effects			Temporary effects		
% Proportion	50–100 (89.92%)	101–200 (6.63%)	>200 (3.45%)	50–100 (89.92%)	101–200 (6.63%)	>200 (3.45%)	50–100 (89.92%)	101–200 (6.63%)	>200 (3.45%)
Variables									
Size	8.40E-06 (8.88E-06)	4.01E-05 (2.90E-05)	2.38E-06 (1.73E-06)	-5.37E-08 (3.82E-06)	2.86E-06 (1.08E-05)	4.81E-06** (2.11E-06)	8.83E-06 (8.56E-06)	3.81E-05 (3.02E-05)	-2.35E-06 (2.10E-06)
Volatility	0.0455 (0.1515)	1.1161*** (0.4051)	-0.2610 (0.2120)	-1.0946*** (0.2200)	1.4354** (0.6454)	-1.0247*** (0.3711)	0.1881 (0.2073)	-0.2271 (0.7360)	-0.0372 (0.4340)
Turnover	-0.1514 (0.4487)	-0.6615 (0.6125)	0.0529 (0.1749)	-0.3513 (0.2838)	-0.3675 (0.3304)	-0.2470** (0.1228)	0.1578 (0.4527)	-0.3157 (0.5964)	0.2986 (0.1260)
Market return	0.0179** (0.0077)	-0.0810 (0.0971)	0.0226* (0.0140)	0.0212*** (0.0080)	-0.0140** (0.0378)	0.0002 (0.0253)	0.0394*** (0.0107)	-0.0089 (0.1006)	0.0221 (0.0284)
Momentum	-0.0043* (0.0026)	-0.0379 (0.0305)	-0.0133* (0.0073)	-0.0066** (0.0033)	0.0385*** (0.0112)	-0.0279*** (0.0117)	0.0028 (0.0040)	-0.0743** (0.0349)	0.0136 (0.0200)
BAS	-0.0547 (0.0529)	-1.1884 (0.9621)	-0.1837 (0.3443)	-0.2725** (0.1193)	-0.4582** (0.2073)	-0.0021 (0.2804)	0.0310 (0.1297)	-0.7479 (0.9649)	-0.1882 (0.4088)
TD_1	-0.0002 (0.0008)	0.0078 (0.0087)	-0.0006 (0.0027)	-0.0005* (0.0003)	-0.0018 (0.0012)	0.0009 (0.0033)	0.0003 (0.0009)	0.0095 (0.0086)	-0.0016 (0.0030)
TD_2	0.0005 (0.0007)	0.0090 (0.0091)	0.0017 (0.0015)	-0.0003 (0.0002)	0.0005 (0.0010)	0.0002 (0.0008)	0.0008 (0.0008)	0.0085 (0.0091)	0.0015 (0.0014)
TD_3	0.0004 (0.0008)	0.0082 (0.0072)	0.0012 (0.0010)	0.0003 (0.0003)	-0.0007 (0.0013)	0.0002 (0.0012)	0.0006 (0.0008)	0.0089 (0.0074)	0.0009 (0.0016)
TD_4	0.0005 (0.0008)	0.0101 (0.0092)	0.0011 (0.0010)	-0.0005* (0.0003)	-0.0007 (0.0011)	9.27E-05 (0.0010)	0.0010 (0.0008)	0.0108 (0.0093)	0.0010 (0.0013)
TD_5	0.0002 (0.0008)	0.0093 (0.0090)	0.0020* (0.0011)	-0.0010*** (0.0003)	-0.0004 (0.0010)	0.0010 (0.0011)	0.0012 (0.0008)	0.0096 (0.0090)	0.0011 (0.0012)

(continued)

Table 4.3 (continued)

TD_6	0.0006	0.0089	0.0014	0.0002	8.21E-05	0.0043**	0.0005	0.0088	−0.0029
	(0.0008)	(0.0084)	(0.0012)	(0.0006)	(0.0011)	(0.0022)	(0.0009)	(0.0085)	(0.0025)
TD_7	0.0006	0.0097	0.0013	−0.0005*	−7.63E-06	0.0002	0.0011	0.0096	0.0011
	(0.0007)	(0.0087)	(0.0013)	(0.0003)	(0.0007)	(0.0010)	(0.0008)	(0.0087)	(0.0016)
TD_8	3.02E-05	0.0089	0.0020**	−0.0005**	0.0005	0.0015	0.0005	0.0084	0.0004
	(0.0007)	(0.0078)	(0.0009)	(0.0002)	(0.0008)	(0.0012)	(0.0007)	(0.0078)	(0.0013)
TD_9	0.0003	0.0075	0.0025*	−0.0004**	−0.0004	−0.0001	0.0007	0.0078	0.0026
	(0.0007)	(0.0070)	(0.0014)	(0.0002)	(0.0008)	(0.0011)	(0.0007)	(0.0071)	(0.0014)
DD_1	0.0006	−0.0114	−0.0020*	−0.0005	0.0001	−0.0024	0.0011	−0.0114	0.0003
	(0.0009)	(0.0108)	(0.0011)	(0.0003)	(0.0017)	(0.0015)	(0.0009)	(0.0110)	(0.0016)
DD_2	0.0006	0.0012	−0.0006	−0.0004*	−2.42E-06	−0.0014	0.0010	0.0012	0.0007
	(0.0009)	(0.0016)	(0.0013)	(0.0002)	(0.0008)	(0.0012)	(0.0009)	(0.0018)	(0.0017)
DD_3	0.0008	0.00154	−0.0016	0.0002	−0.0004	−0.0019	0.0007	0.0019	0.0002
	(0.0009)	(0.0013)	(0.0013)	(0.0003)	(0.0008)	(0.0014)	(0.0009)	(0.0015)	(0.0020)
DD_4	0.0005	0.0003	−0.0014	−0.0003	−0.0012	−0.0023*	0.0008	0.0015	0.0008
	(0.0009)	(0.0012)	(0.0012)	(0.0002)	(0.0008)	(0.0012)	(0.0009)	(0.0014)	(0.0016)
MD_1	0.0003	0.0002	0.0015	0.0002	−0.0015	0.0010	7.93E-05	0.0016	0.0005
	(0.0006)	(0.0021)	(0.0010)	(0.0004)	(0.0011)	(0.0007)	(0.0006)	(0.0021)	(0.0011)
MD_2	0.0004	0.0023	−0.0006	0.0006	0.0024	0.0021	−0.0001	0.0002	−0.0026
	(0.0005)	(0.0026)	(0.0009)	(0.0005)	(0.0022)	(0.0018)	(0.0006)	(0.0032)	(0.0023)
MD_3	−7.77E-05	0.0042	0.0010	−6.88E-05	0.0004	−0.0010	−1.04E-06	0.0038	0.0020
	(0.0005)	(0.0029)	(0.0007)	(0.0004)	(0.0010)	(0.0012)	(0.0005)	(0.0028)	(0.0015)
MD_4	−0.0009	0.0021	0.0007	−7.69E-05	0.0001	0.0014	−0.0008	0.0021	−0.0006
	(0.0014)	(0.0022)	(0.0008)	(0.0004)	(0.0011)	(0.0010)	(0.0014)	(0.0021)	(0.0010)
MD_5	−0.0004	−0.0008	−0.0018	−0.0002	0.0003	−0.0005	−0.0002	−0.0010	−0.0013
	(0.0005)	(0.0021)	(0.0012)	(0.0004)	(0.0011)	(0.0010)	(0.0005)	(0.0021)	(0.0019)
MD_6	0.0004	0.0011	−0.0005	0.0003	−9.19E-07	0.0011	0.0001	0.0011	−0.0016
	(0.0005)	(0.0024)	(0.0010)	(0.0004)	(0.0010)	(0.0016)	(0.0005)	(0.0025)	(0.0014)
MD_7	6.23E-05	0.0012	0.0006	−0.0002	0.0011	−0.0015	0.0003	0.0002	0.0021
	(0.0005)	(0.0019)	(0.0014)	(0.0004)	(0.0011)	(0.0026)	(0.0006)	(0.0020)	(0.0019)
MD_8	0.0004	0.0024	0.0035***	0.0003	−0.0011	0.0004	2.14E-05	0.0035	0.0032
	(0.0005)	(0.0028)	(0.0013)	(0.0004)	(0.0011)	(0.0014)	(0.0006)	(0.0026)	(0.0021)

MD$_9$	0.0004	0.0017	8.08E-05	0.0003	0.0025	0.0003	0.0015	-0.0013
	(0.0005)	(0.0021)	(0.0005)	(0.0018)	(0.0020)	(0.0006)	(0.0030)	(0.0017)
MD$_{10}$	0.0002	-0.0014	2.79E-05	-0.0010	-0.0007	0.0002	-0.0003	0.0013
	(0.0005)	(0.0028)	(0.0004)	(0.0014)	(0.0012)	(0.0005)	(0.0030)	(0.0015)
MD$_{11}$	0.0001	-0.0147	0.0001	0.0004	-0.0005	-3.26E-05	-0.0151	-0.0006
	(0.0005)	(0.0159)	(0.0004)	(0.0013)	(0.0007)	(0.0005)	(0.0161)	(0.0011)
Constant	-0.0018	-0.0136	0.0007	-0.0013	0.0020	-0.0023	-0.0126	-0.0013
	(0.0017)	(0.0095)	(0.0006)	(0.0012)	(0.0020)	(0.0018)	(0.0096)	(0.0024)
Observations	7514	554	7514	554	288	7514	554	288
R^2	0.0027	0.0363	0.0232	0.2208	0.2484	0.0066	0.0359	0.1901

The table reports regression results for all purchase block trades of December maturity EUA futures contracts executed on the ECX platform between January 2008 and April 2011. The coefficients are reported along with the standard errors (in parenthesis). The following regression is estimated using OLS with Newey and West (1987) heteroscedastic and autocorrelation consistent covariance matrix:

$$PI_t = \gamma_0 + \gamma_x X_t + \gamma_2 \sum_{i=1}^{9} TD_i + \gamma_3 \sum_{i=1}^{4} DD_i + \gamma_4 \sum_{i=1}^{11} MD_i + \varepsilon_t$$

where PI_t corresponds to one of three price impact measures: total price impact, permanent price impact and temporary price impact. The explanatory variables are computed as follows. X_t is a vector of six explanatory variables (size, volatility, turnover, market return, momentum and BAS) defined below. TD_i, DD_i and MD_i are dummy variables for time (hour) of day, day of week and month of year and are further defined below. Size represents the natural logarithm of the number of December maturity futures contracts for each block trade; volatility is the standard deviation of trade to trade returns prior to the block trade on the trading day; turnover is the natural logarithm of the futures contracts turnover on the trading day prior to the block trade, turnover is the ratio of total trade volume prior to the block to the prevailing open interest estimates; market return is the return of EUA Futures contract specific index computed by the ECX; momentum corresponds to the cumulative return on the specific EUA Futures contract in the five days prior to the block trade; BAS is the prevailing relative bid-ask spread at the time the block trade is executed, BAS is measured as the last ask price prior to the block trade minus the last bid price before the block trade, divided by the average of both prices. TD_1 to TD_9 equal one if the block trade occurs in any of the corresponding hour of trade from the first hour (1) to the ninth hour (9) of the trading day and 0 otherwise. DD_1 to DD_4 equal 1 if the block trade occurs in any of the corresponding day of the week from Monday (1) to Thursday (4) of the trading week and 0 otherwise. Any of MD_1 to MD_{11} equals 1 if the block trade occurs in any of the corresponding months of the year from January (1) to November (11) and 0 otherwise. One EUA Futures contract has an underlying of 1000 EUAs

***, ** and * indicate statistical significance at 1, 5 and 10% level, respectively

for momentum are statistically significant for all three sizes; again the results suggest that mid-size blocks induce higher price impact on a price run-up, as reported by Frino et al. (2007), while results on the other two sizes confirm the argument of Saar (2001) and results in Table 4.2. The negative and statistically significant coefficients confirm that less price impact is induced on a price run-up.

Differentiation (outright dominance) of the mid-size blocks group from (over) the largest blocks category is consistent with the hypothesis and evidence presented by Barclay and Warner (1993) after testing block trade price impacts on NYSE stock prices. In their sample, most of the cumulative stock price change is due to medium-sized trades. Here, the interesting pattern evolving may be an indication of 100,000 EUAs becoming the threshold for price impact effects. Already, we show that the small blocks induce more temporary than permanent price impact, thus implying that most of the trades in this group are liquidity seeking. This, coupled with the large volume of small blocks of 50,000 EUAs, suggests a gradual erosion of the market's view of a 50,000 EUA-worth trade as a block trade. The ECX sets the standard for what is regarded as a block trade, and currently, it stands at 50,000 EUAs, according to exchange rules. Markets have, however, been known to induce structural shifts in response to emerging trading *culture*. The ECX, as an EU-ETS platform, is a product of political action and may not be subject to the same expectations as regular markets developed as engines of wealth creation. Even still, the market seems to be gradually taking on a life of its own. This, however, does not explain why the largest block category shows fewer price impact effects than the mid-size category. Barclay and Warner (1993) argue that under certain conditions, informed traders, rather than trading in large sizes, would usually split up their trades into smaller chunks that fall into the medium-sized category, hence the asymmetric phenomenon. In addition, we suggest the following: the frequency of on-screen purchase block trades >200,000 EUAs on the ECX (3.45%) over three years and four months is very low. The low frequency levels may be a contributing factor to the low coefficient estimates. Perhaps infrequent trade sizes are likely to induce less price reaction than those that are fairly regular. Furthermore, in our dataset, more than 98% of the large blocks occur outside of the first hour of the normal trading day. The total effects coefficient for the mid-size category indicates that greater price impact occurs during the first hour of the trading day for this group of trades. Therefore, since trades during the other periods in the day are less likely to induce price impact than those executed during the first hour, the effects of the large blocks may have been consequently muted by a general reduced price reaction to such trades during the other periods of the normal trading day. The R^2 values for the equation estimations range from 2.32 to 24.84% for total effects estimates, which is an indication of the significant explanatory power of the model, especially for the large block trades. This further evidences their informativeness.

Table 4.4 shows the results for sell block trades. We observe trends similar to those in Table 4.3, with approximately 91% of executed trades in the sample made up of the small blocks. Also, as in the purchase estimates, there is a dearth of coefficients that are significantly different from zero. Like the purchase block results, the results in Table 4.4 are consistent with sell blocks estimates in Table 4.2. However, unlike with the purchase blocks, the small sell blocks contribute more to the direction of price impact observed in Table 4.2 than the larger blocks. The only exception seems to be for temporary effects due to volatility, which is principally driven by the mid-size blocks. The negative and statistically significant value of the mid-size blocks show that increased volatility results in higher price impact for mid-size blocks. The opposite is the case for the small and large blocks. Consistent with Alzahrani, Gregoriou and Hudson (2013), the significant total effects coefficient for the small blocks is larger than for the larger sizes. This indicates that smaller sized sell block trades are more informative than larger ones. Professional traders have long been known to split large block trades into smaller trades to avoid early detection of their information content (see Barclay and Warner 1993; Chakravarty 2001). Although microstructure studies show that informed trades are discernible also from the direction and frequency of trades, trades fragmentation potentially mutes the price impact of block trades (Keim and Madhavan 1996). The permanent effects coefficients for market return seem to support this view; thus, there are larger price impacts for the small sell blocks, as they are perceived as being more informative.

Table 4.4 Determinants of price impact and block trade sizes (sales)

	Permanent effects			Total effects			Temporary effects		
% Proportion	50–100 (90.85%)	101–200 (5.91%)	>200 (3.24%)	50–100 (90.85%)	101–200 (5.91%)	>200 (3.24%)	50–100 (90.85%)	101–200 (5.91%)	>200 (3.24%)
Variables									
Size	−1.62E−05*** (4.32E−06)	1.36E−06 (5.83E−06)	−3.91E−07 (1.41E−06)	−4.26E−06** (2.04E−06)	−1.23E−07 (3.09E−06)	−2.09E−06** (9.42E−07)	−1.99E−06 (3.48E−06)	1.50E−06 (4.36E−06)	1.71E−06 (1.22E−06)
Volatility	0.0426 (0.1167)	0.1953 (0.2846)	0.1486 (0.2211)	0.2091** (0.0920)	0.4329** (0.2094)	−0.4295* (0.2576)	0.0330 (0.0698)	−0.2366*** (0.0819)	0.5814** (0.2700)
Turnover	−0.0020 (0.0033)	−0.0099 (0.0037)	0.0010 (0.0116)	0.0668 (0.1994)	−0.1157 (0.2193)	0.0682 (0.0525)	−0.2682*** (0.0432)	0.0166 (0.2173)	−0.0680 (0.1013)
Market return	0.0182*** (0.0058)	0.0158 (0.0119)	0.0083 (0.0096)	0.0125*** (0.0033)	0.0059 (0.0076)	0.0247*** (0.0060)	0.0057 (0.0053)	0.0099 (0.0086)	−0.0165* (0.0095)
Momentum	−0.0013 (0.0025)	−0.0063 (0.0056)	−0.0222*** (0.0062)	−0.0012 (0.0013)	−0.0012 (0.0027)	−0.0111*** (0.0038)	−9.88E−06 (0.0021)	−0.0051 (0.0047)	−0.0111** (0.0048)
BAS	0.2873** (0.1394)	−0.3524*** (0.1290)	0.2609 (0.3387)	0.1548 (0.1082)	−0.3334*** (0.0935)	0.1578 (0.3322)	0.0314 (0.0810)	−0.0203 (0.1322)	0.1044 (0.1159)
TD_1	0.0003 (0.0003)	0.0007 (0.0005)	−0.0018 (0.0012)	0.0003* (0.0002)	0.0003 (0.0004)	−0.0018* (0.0010)	−1.39E−05 (0.0003)	0.0004 (0.0003)	−7.10E−06 (0.0007)
TD_2	−0.0001 (0.0003)	0.0008 (0.0006)	−0.0003 (0.0009)	−0.0001 (0.0002)	0.0007 (0.0005)	−0.0004 (0.0006)	2.58E−05 (0.0002)	6.38E−05 (0.0004)	8.66E−05 (0.0009)
TD_3	0.0002 (0.0003)	0.0005 (0.0008)	−0.0007 (0.0008)	9.14E−05 (0.0002)	0.0002 (0.0004)	−0.0008** (0.0004)	7.98E−05 (0.0002)	0.0003 (0.0006)	9.35E−05 (0.0006)
TD_4	−7.15E−05 (0.0003)	0.0007 (0.0006)	0.0012* (0.0007)	5.55E−05 (0.0002)	0.0003 (0.0004)	0.0004 (0.0005)	−0.0001 (0.0002)	0.0004 (0.0004)	0.0008 (0.0005)
TD_5	−0.0006** (0.0003)	0.0004 (0.0006)	0.0004 (0.0007)	−0.0003** (0.0002)	0.0004 (0.0003)	−8.39E−05 (0.0004)	−0.0003 (0.0002)	−2.58E−05 (0.0005)	0.0004 (0.0006)
TD_6	0.0004 (0.0003)	0.0002 (0.0006)	0.0014** (0.0007)	0.0003* (0.0002)	0.0002 (0.0004)	0.0003 (0.0003)	6.38E−05 (0.0002)	7.78E−06 (0.0005)	0.0011* (0.0006)

	(1)	(2)	(3)	(4)	(5)	(6)	(7)	(8)	(9)
TD_7	8.43E-05 (0.0003)	-0.0001 (0.0005)	0.0016** (0.0008)	3.77E-05 (0.0001)	0.00012 (0.0004)	0.0004 (0.0004)	4.75E-05 (0.0002)	-0.0003 (0.0004)	0.0012 (0.0008)
TD_8	-0.0005** (0.0003)	-0.0006 (0.0006)	0.0013 (0.0009)	-0.0003** (0.0001)	-0.0003 (0.0003)	0.0003 (0.0006)	-0.0002 (0.0002)	-0.0003 (0.0005)	0.0011* (0.0006)
TD_9	-0.0003 (0.0003)	0.0008 (0.0005)	0.0020** (0.0008)	-0.0003** (0.0001)	0.0001 (0.0004)	0.0003 (0.0006)	-1.59E-05 (0.0002)	0.0007* (0.0004)	0.0017*** (0.0006)
DD_1	0.0003 (0.0003)	-0.0001 (0.0005)	5.54E-05 (0.0007)	4.61E-05 (0.0001)	0.0002 (0.0004)	-0.0011** (0.0005)	0.0002 (0.0002)	-0.0003 (0.0004)	0.0011* (0.0006)
DD_2	0.0001 (0.0003)	-0.0013*** (0.0004)	6.51E-05 (0.0006)	-8.76E-05 (0.0001)	-0.0003 (0.0003)	-0.0010** (0.0004)	0.0002 (0.0002)	-0.0010 (0.0003)	0.0010* (0.0006)
DD_3	0.0001 (0.0002)	3.57E-05 (0.0005)	-9.30E-05 (0.0007)	0.0002 (0.0001)	0.0002 (0.0003)	-0.0006 (0.0005)	-3.55E-05 (0.0002)	-0.0002 (0.0004)	0.0005 (0.0005)
DD_4	-0.0002 (0.0002)	-0.0010** (0.0004)	-0.0006 (0.0007)	2.06E-05 (0.0001)	-0.0005 (0.0003)	-0.0011*** (0.0004)	-0.0002 (0.0002)	-0.0005 (0.0004)	0.0005 (0.0006)
MD_1	0.0006* (0.0004)	0.0007 (0.0006)	0.0030** (0.0013)	0.0008*** (0.0002)	0.0002 (0.0004)	0.0015** (0.0007)	-0.0002 (0.0002)	0.0004 (0.0004)	0.0015* (0.0009)
MD_2	0.0009** (0.0004)	3.06E-05 (0.0006)	0.0032*** (0.0012)	0.0007*** (0.0002)	0.0003 (0.0004)	0.0009 (0.0006)	0.0003 (0.0003)	-0.0003 (0.0004)	0.0023*** (0.0008)
MD_3	0.0007* (0.0003)	5.19E-05 (0.0006)	0.0036*** (0.0010)	0.0005** (0.0002)	2.42E-05 (0.0004)	0.0016** (0.0006)	0.0002 (0.0003)	2.68E-05 (0.0004)	0.0020*** (0.0007)
MD_4	0.0006* (0.0003)	0.0008 (0.0008)	0.0029*** (0.0011)	0.0006*** (0.0002)	0.0007 (0.0005)	0.0019*** (0.0007)	1.87E-05 (0.0003)	0.0001 (0.0005)	0.0010 (0.0007)
MD_5	0.0003 (0.0004)	-0.0001 (0.0006)	0.0023*** (0.0010)	0.0005** (0.0003)	-0.0003 (0.0003)	0.0023*** (0.0007)	-0.0002 (0.0003)	0.0001 (0.0004)	9.29E-05 (0.0006)
MD_6	0.0010** (0.0004)	0.0006 (0.0006)	0.0024* (0.0013)	0.0008*** (0.0003)	0.0004 (0.0004)	0.0012 (0.0008)	0.0002 (0.0003)	0.0002 (0.0005)	0.0011* (0.0007)
MD_7	0.0013*** (0.0004)	-1.44E-05 (0.0007)	0.0023* (0.0013)	0.0008*** (0.0003)	3.80E-05 (0.0004)	0.0016** (0.0008)	0.0005 (0.0003)	-5.16E-05 (0.0006)	0.0006 (0.0009)
MD_8	0.0002 (0.0004)	0.0009 (0.0008)	0.0054*** (0.0016)	0.0005** (0.0002)	0.0009 (0.0006)	0.0027*** (0.0008)	-0.0003 (0.0004)	5.79E-05 (0.0005)	0.0027*** (0.0010)
MD_9	0.0005 (0.0003)	-0.0005 (0.0013)	0.0029*** (0.0010)	0.0007*** (0.0002)	-0.0003 (0.0004)	0.0007 (0.0007)	-0.0002 (0.0003)	-0.0001 (0.0010)	0.0022*** (0.0006)

(continued)

Table 4.4 (continued)

MD_{10}	0.0005	−0.0009	0.0026***	0.0005*	−0.0003	0.0010**	4.53E-05	−0.0006	0.0016***
	(0.0005)	(0.0010)	(0.0009)	(0.0003)	(0.0005)	(0.0006)	(0.0003)	(0.0007)	(0.0006)
MD_{11}	0.0005	0.0001	0.0031***	0.0005**	0.0001	0.0009*	−3.50E-05	−4.21E-06	0.0022***
	(0.0003)	(0.0006)	(0.0010)	(0.0002)	(0.0004)	(0.0006)	(0.0003)	(0.0004)	(0.0006)
Constant	0.0003	0.0001	−0.0034***	−6.20E-05	−7.09E-05	0.0005	0.0003	0.0002	−0.0039***
	(0.0005)	(0.0010)	(0.0011)	(0.0003)	(0.0006)	(0.0007)	(0.0004)	(0.0007)	(0.0008)
Observations	7594	494	271	7594	494	271	7594	494	271
R^2	0.0228	0.0857	0.2089	0.0289	0.1342	0.2517	0.0102	0.0613	0.2467

The table reports regression results for sell block trades of December maturity EUA Futures contracts executed on the ECX platform between January 2008 and April 2011. The coefficients are reported along with the standard errors (in parenthesis). The following regression is estimated using OLS with Newey and West (1987) heteroscedastic and autocorrelation consistent covariance matrix:

$$PI_t = \gamma_0 + \gamma_x X_t + \gamma_2 \sum_{i=1}^{9} TD_i + \gamma_3 \sum_{i=1}^{4} DD_i + \gamma_4 \sum_{i=1}^{11} MD_i + \varepsilon_t$$

where PI_t corresponds to one of three price impact measures: total price impact, permanent price impact and temporary price impact. The explanatory variables are computed as follows. X_t is a vector of six explanatory variables (size, volatility, turnover, market return, momentum and BAS) defined below. TD_i, DD_i and MD_i are dummy variables for time (hour) of day, day of week and month of year and are further defined below. Size represents the natural logarithm of the number of December maturity futures contracts for each block trade; volatility is the standard deviation of trade to trade returns prior to the block trade on the trading day; Turnover is the natural logarithm of the futures contracts turnover on the trading day prior to the block trade, turnover is the ratio of total trade volume prior to the block to the prevailing open interest estimates; market return is the return of EUA Futures contract specific index computed by the ECX; momentum corresponds to the cumulative return on the specific EUA Futures contract in the five days prior to the block trade; BAS is the prevailing relative bid-ask spread at the time the block trade is executed, BAS is measured as the last ask price prior to the block trade minus the last bid price before the block trade, divided by the average of both prices. TD_{i1} to TD_{i9} equal 1 if the block trade occurs in any of the corresponding hour of trade from the first hour (1) to the ninth hour (9) of the trading day and 0 otherwise. DD_{i1} to DD_{i4} equal 1 if the block trade occurs in any of the corresponding day of the week from Monday (1) to Thursday (4) of the trading week and 0 otherwise. Any of MD_{i1} to MD_{i11} equals 1 if the block trade occurs in any of the corresponding months of the year from January (1) to November (11) and 0 otherwise. One EUA Futures contract has an underlying of 1000 EUAs

***, ** and * indicate statistical significance at 1, 5 and 10% level, respectively

Another interesting set of results to note in Table 4.4 is the fact that momentum coefficients are negative all through, consistent with Table 4.2. When viewed in tandem with the block purchase estimates in previous tables, we can argue that block purchase trades behaviour on the ECX is largely consistent with Saar (2001) (cumulative lagged returns reduce price impact) and that of sell block trades with Frino et al. (2007) (larger price run-ups lead to larger price impact). The time-related dummies results are qualitatively similar to sell blocks estimates in Table 4.2. The R^2 values are generally larger than for previous regressions, with the model being more fitted to explaining price impact for the large sell blocks; the R^2 values range from 20.89 to 25.17% for the largest sized sell blocks. The general trend exhibited in Table 4.4 suggests that sell block trades executed on the ECX on the whole are less likely to induce price impact than purchase block trades.

4.5 Chapter Summary

The key finding of this study is that the price impact of block trades in emissions permit markets is largely different from the price impact of block trades in conventional financial markets, with emissions permit markets experiencing generally lower price impact than traditional financial markets. Block trades increasingly constitute large Euro volumes of trades in the EU-ETS, as more installations try to avoid counterparty risks by trading on platforms rather than OTC; in any case, most OTC trades are registered on exchanges to avoid such risks. Our study therefore is of significance to CFI traders and exchange operators alike. By examining the determinants of block trades, we improve understanding of the impact of larger than regular trades on an environmental platform. This understanding is useful for regulators and exchange operators, and it will contribute to the development of market design considerations. For example, for purchase blocks, we find little evidence of large impact for ECX block trades within the threshold of 50,000–100,000 EUAs and above the threshold of 200,000 EUAs; instead, the impact is stronger for the mid-size blocks between 100,000 and 200,000 EUAs. This suggests a disparity between the platform operator's expectation of price and trading dynamics, and the view of the market participants.

Results show that most of the block trades on the ECX occur at the minimum quantity for the exchange-sanctioned block trade size of 50,000 EUAs. This is consistent for both buyer- and seller-initiated block trades; it also indicates that traders on either side of block trades on the ECX employ identical trading tactics in terms of order placing. The evidence suggests that stealth trading is a strategy being employed by most block traders on the platform. The low volume of block trades—16,715 (1.74%) out of a total of 961,131 trades in our final sample—also suggests that hitherto block trading intentions may have been executed by splitting aggregate large orders into traded quantities below the block trade threshold of 50,000 EUAs. This suggestion is reinforced by the nature of the EU-ETS, where most participants are either big compliance traders or institutional investors, who are expected to be trading in large quantities. In comparison with conventional instruments, the price impact of carbon futures on the ECX is small and largely statistically insignificant. Lack of price reaction to large trades can be viewed as a possible consequence of thin trading (see Ball and Finn 1989). Although trading has advanced in the EU-ETS, it remains very low in comparison to established markets. Since there is little price reaction, there is very little opportunity to benefit from price shocks. We find some proof of price impact asymmetry for buyer- and seller-initiated block trades. Some results also suggest that sellers pay a premium, rather than buyers, on buyer-initiated trades, which clearly contradicts many studies. However, it is not surprising that sellers rather than buyers pay a premium on the ECX when the market structure is considered. The ECX is a derivatives exchange for emission permits, which are required for submission only once a year; hence, for most trading days the permits are largely hedging instruments. Compliance buyers do not need to hold on to the underlying instruments all year round and therefore only need to take long positions in the market to ensure that they are insulated against penalties for non-compliance. In the event that they are in possession of excess instruments, they can undertake a short position. Considering that the permits hold little value to a compliance trader unless they are being submitted, many may make concessions to sell them even when the trade is buyer-initiated, and such a realisation may trigger the buyer's interest in the first place.

This chapter's study also provides evidence that lower price impact is characterised by wider spreads. For buyer-initiated block orders, trade execution induces larger price impact in the ECX during the middle of the trading day than during the first and last trading hours. We find evidence in support of positive price run-up leading to both lower price impact and higher price impact, depending on the trade sign. For block purchases, there is smaller price impact when a trade occurs after a price run-up; for block sales, there is greater price impact. There is, however, also a block trade size dependency to this, as shown in Sect. 4.4. In many cases, the most information-laden trades are not the largest ones, but the medium (for most purchases) and small (for most sells) trades. Policymakers must therefore ensure that regulations in the emissions markets keep pace with trading innovations. Our findings also have implications for the participants in the EU-ETS who must trade in emissions instruments for the purpose of compliance. This class of participants is more likely to trade in large blocks; thus, the findings here can inform their trading strategies. Specifically, the dissimilarities to conventional platforms that we document in this chapter underscore the need for trading re-orientation to attain a required level of trading sophistication in the market.

Notes

1. Upstairs trading refers to transactions facilitated by brokers and broker-dealers, such trades are generally hidden from most of the market. Interested parties inform their brokers/broker-dealers of their orders and the brokers/broker-dealers in turn shop for counterparties to take the other side of such orders. The downstairs market is the central limit order book (CLOB), where all orders are submitted anonymously, but are displayed until filled or cancelled.
2. The tick rule (Lee and Ready 1991) is also employed with very similar trade sign classification portioning.
3. For robustness, we consider analysis based on volume-weighted price impact measures in order to eliminate some of the noise in the data, if indeed this was prominent. We find that the results are quantitatively similar. The major reason for this is that approximately 90% of the block trades are in the 50,000–100,000 EUAs size category, thus the trades mainly have a similar multiplying factor.

4. We also use the plain number of instruments traded, the natural logarithm of the Euro value of the block trades, the instrument volume relative to the average daily instrument volume, and the Euro value relative to the average daily trading Euro value. As is the case with Frino et al. (2007), the natural logarithm of the number of instruments traded provides the best fit.
5. We also use standard deviation of the execution price of trades, in line with Alzahrani et al. (2013). The results are quantitatively similar.
6. We also use open interest on its own as a measure of liquidity; however, the results obtained are less significant across all the models.
7. The expectation is for higher levels of volatility to lead to higher block trade price impact. Thus, block purchases will induce positively skewed price impact and sell block trades will induce negative price impact when the market becomes relatively more volatile.
8. We also divide the trading day into three intervals, in line with previous studies, in order to test the intraday period dependencies. The results are qualitatively similar.

References

Aitken, M., & Frino, A. (1996). Execution Costs Associated with Institutional Trades on the Australian Stock Exchange. *Pacific-Basin Finance Journal, 4*, 45–58.

Alzahrani, A. A., Gregoriou, A., & Hudson, R. (2013). Price Impact of Block Trades in the Saudi Stock Market. *Journal of International Financial Markets, Institutions and Money, 23*, 322–341.

Ball, R., & Finn, F. J. (1989). The Effect of Block Transactions on Share Prices: Australian Evidence. *Journal of Banking & Finance, 13*, 397–419.

Barclay, M. J., & Warner, J. B. (1993). Stealth Trading and Volatility: Which Trades Move Prices? *Journal of Financial Economics, 34*, 281–305.

Bessembinder, H., & Seguin, P. J. (1992). Futures-Trading Activity and Stock Price Volatility. *The Journal of Finance, 47*, 2015–2034.

Chakravarty, S. (2001). Stealth-Trading: Which Traders' Trades Move Stock Prices? *Journal of Financial Economics, 61*, 289–307.

Chan, L. K. C., & Lakonishok, J. (1993). Institutional Trades and Intraday Stock Price Behavior. *Journal of Financial Economics, 33*, 173–199.

Chan, L. K. C., & Lakonishok, J. (1995). The Behavior of Stock Prices around Institutional Trades. *The Journal of Finance, 50*, 1147–1174.

Chiyachantana, C. N., Jain, P. K., Jiang, C., & Wood, R. A. (2004). International Evidence on Institutional Trading Behavior and Price Impact. *The Journal of Finance, 59*, 869–898.

Chordia, T., Roll, R., & Subrahmanyam, A. (2008). Liquidity and Market Efficiency. *Journal of Financial Economics, 87*, 249–268.

Chou, R. K., Wang, G. H. K., Wang, Y.-Y., & Bjursell, J. (2011). The Impacts of Large Trades by Trader Types on Intraday Futures Prices: Evidence from the Taiwan Futures Exchange. *Pacific-Basin Finance Journal, 19*, 41–70.

Conrad, J. S., Johnson, K. M., & Wahal, S. (2001). Institutional Trading and Soft Dollars. *The Journal of Finance, 56*, 397–416.

Domowitz, I., Glen, J., & Madhavan, A. (2001). Liquidity, Volatility and Equity Trading Costs across Countries and over Time. *International Finance, 4*, 221–255.

Easley, D., Hvidkjaer, S., & O'Hara, M. (2002). Is Information Risk a Determinant of Asset Returns? *The Journal of Finance, 57*, 2185–2221.

Easley, D., & O'Hara, M. (1987). Price, Trade Size, and Information in Securities Markets. *Journal of Financial Economics, 19*, 69–90.

Frino, A., Jarnecic, E., & Lepone, A. (2007). The Determinants of the Price Impact of Block Trades: Further Evidence. *Abacus, 43*, 94–106.

Fung, H.-G., & Patterson, G. A. (1999). The Dynamic Relationship of Volatility, Volume, and Market Depth in Currency Futures Markets. *Journal of International Financial Markets, Institutions and Money, 9*, 33–59.

Gemmill, G. (1996). Transparency and Liquidity: A Study of Block Trades on the London Stock Exchange under Different Publication Rules. *The Journal of Finance, 51*, 1765–1790.

Glosten, L. R., & Harris, L. E. (1988). Estimating the Components of the Bid/Ask Spread. *Journal of Financial Economics, 21*, 123–142.

Gregoriou, A. (2008). The Asymmetry of the Price Impact of Block Trades and the Bid-Ask Spread. *Journal of Economic Studies, 35*, 191–199.

Holthausen, R. W., Leftwich, R. W., & Mayers, D. (1987). The Effect of Large Block Transactions on Security Prices: A Cross-Sectional Analysis. *Journal of Financial Economics, 19*, 237–267.

Holthausen, R. W., Leftwich, R. W., & Mayers, D. (1990). Large-Block Transactions, the Speed of Response, and Temporary and Permanent Stock-Price Effects. *Journal of Financial Economics, 26*, 71–95.

Hu, S. (1997). *Trading Turnover and Expected Stock Returns: The Trading Frequency Hypothesis and Evidence from the Tokyo Stock Exchange*. National Taiwan University Working Paper, Taipei.

Ibikunle, G., Gregoriou, A., & Pandit, N. R. (2016). Price Impact of Block Trades: The Curious Case of Downstairs Trading in the EU Emissions Futures Market. *The European Journal of Finance, 22*, 120–142.

Keim, D., & Madhavan, A. (1996). The Upstairs Market for Large-Block Transactions: Analysis and Measurement of Price Effects. *The Review of Financial Studies, 9*, 1–36.

Kossoy, A., & Ambrosi, P. (2010). *State and Trends of the Carbon Markets, 2010.* The World Bank Report, Washington, DC.

Kraus, A., & Stoll, H. R. (1972). Price Impacts of Block Trading on the New York Stock Exchange. *The Journal of Finance, 27*, 569–588.

Lakonishok, J., & Lev, B. (1987). Stock Splits and Stock Dividends: Why, Who, and When. *The Journal of Finance, 42*, 913–932.

Lee, C. M., & Ready, M. J. (1991). Inferring Trade Direction from Intraday Data. *The Journal of Finance, 46*, 733–746.

Madhavan, A., & Cheng, M. (1997). In Search of Liquidity: Block Trades in the Upstairs and Downstairs Markets. *The Review of Financial Studies, 10*, 175–203.

Madhavan, A., Richardson, M., & Roomans, M. (1997). Why Do Security Prices Change? A Transaction-Level Analysis of NYSE Stocks. *The Review of Financial Studies, 10*, 1035–1064.

Mizrach, B., & Otsubo, Y. (2014). The Market Microstructure of the European Climate Exchange. *Journal of Banking & Finance, 39*, 107–116.

Montagnoli, A., & de Vries, F. P. (2010). Carbon Trading Thickness and Market Efficiency. *Energy Economics, 32*, 1331–1336.

Newey, W. K., & West, K. D. (1987). A Simple, Positive Semi-definite, Heteroskedasticity and Autocorrelation Consistent Covariance Matrix. *Econometrica, 55*, 703–708.

Rittler, D. (2012). Price Discovery and Volatility Spillovers in the European Union Emissions Trading Scheme: A High-Frequency Analysis. *Journal of Banking & Finance, 36*, 774–785.

Rotfuß, W. (2009). *Intraday Price Formation and Volatility in the European Union Emissions Trading Scheme.* Centre for European Economic Research (ZEW) Working Paper, Manheim.

Saar, G. (2001). Price Impact Asymmetry of Block Trades: An Institutional Trading Explanation. *The Review of Financial Studies, 14*, 1153–1181.

Sarr, A., & Lybek, T. (2002). *Measuring Liquidity in Financial Markets.* International Monetary Fund Working Paper WP/02/232, Washington, DC.

5

The Liquidity Effects of Trading Carbon Financial Instruments

5.1 Introduction

Butzengeiger et al. (2001) identify liquidity as a precondition for the success of an ETS. An ETS should involve a sufficient pool of participants in order to ensure that an adequate volume of transactions occurs on a regular basis; this results in the emergence of an explicit price signal to the market. Fragmentation of platforms may inhibit the advancement of liquidity especially in nascent markets like the global carbon market; potentially, liquidity risk resulting from fragmentation of trading platforms is a risk in the EU-ETS as evidenced by low trading volumes during the Phase I of the EU-ETS (see Hill et al. 2008). Liquidity is one of the most vital features investors search for in any financial market before trading. Specifically, financial markets participants usually view a market as liquid if it is possible to transact in the market's instruments with reasonable speed, irrespective of the volume and size of transactions, and with little or no price impact on the instrument. Therefore, trading in liquid securities offer low transaction costs, ease of transactions' settlement and virtually no asymmetric information costs (Sarr and Lybek 2002). For new markets, low transaction costs are vital for the advancement of trading volumes, which are necessary for efficient price discovery (see Chap. 3) Given this importance of liquidity to the success of the EU-ETS and perhaps the establishment of a global cap and trade scheme, in this chapter, we examine the liquidity effects on specific futures contracts trading on the EEX, a significantly lower trading volume EU-ETS platform than the ECX. Thus, if we are to find evidence of liquidity effects, it would most likely occur on a platform like the EEX. We test for liquidity effects by investigating the effects of four key events during the second phase of the EU-ETS.

Liquidity cannot be directly inferred from improvement in transaction volumes in financial markets. Recent empirical contributions have shown that improvements in trading volumes are not necessarily associated with enhanced liquidity. Indeed volume and liquidity are weakly related over time (Johnson 2008). Early studies on the subject of the relatedness of market liquidity and volume include the studies of Foster and Viswanathan (1993) and Lee et al. (1993) both of which show negative correlations. More recently, the use of Dow-Jones 30 industrial stocks and other instruments evidence only insignificant changes in bid-ask spread in relation to turnover (see Jones 2002; Fujimoto 2004). Danielsson and Payne (2010) using order entry rates and depth measures as proxies for liquidity also find a negative correlation between trade volumes and liquidity.

In the first phase (Phase I) of the EU-ETS, there was an overallocation of emission/carbon permits. This, coupled with a series of restrictive regulations, contributed to the occurrence of a largely illiquid market. It can be argued that part of the problem with market quality during the phase is the fact that Phase I was just a trial phase. Indeed, events during the phase gave indications of what could go wrong during the subsequent phases, especially with regards to Phase II, which was the legal Kyoto commitment period. In response to this and by earlier design intentions, there are regulatory changes between Phase I and Phase II (see Table 2.1 in Chap. 2 for an outline). Tighter caps were also introduced. If the implementation of new regulations and tighter caps for Phase II has been successful in reducing the overallocation of carbon allowances and improving market quality, one would expect a significant increase in market liquidity in the EU-ETS overtime during that phase of the scheme. We argue in the introduction to this book that the two main functions of a financial market are price discovery and liquidity and that these two give indication of market quality as well as market efficiency. This chapter therefore investigates whether or not there have been significant improvements in market liquidity since the start of trading in Phase II of the EU-ETS. We would expect that the lessons learned during the first phase would have helped regulators in designing new rules aimed at improving market liquidity. The investigations are carried out using the trading data for the Dec-2008 EEX carbon futures contract, which accounts for more than 70% of traded volumes on the EEX during the period under investigation. The Dec-2008 contract started trading in Phase I, with its underlying permits also issued in Phase I; however, it can only be submitted for compliance towards Phase II emissions. Thus, it is a unique instrument, which allows for the investigation of phase transition effects on liquidity. In the study we present in this chapter, we also extend our investigations to capture the liquidity effects of emissions verification results (for 2007 and 2008 compliance years) and the introduction of the European Commission Regulation (EC) No 994/2008 of 8 October 2008.

The EEX platform is chosen because this study investigates liquidity improvements. The carbon trading division of the EEX was clearly one of the more illiquid platforms during the trial phase. During the early stages of Phase I, there were no trades for significant periods and trading spreads were large. As pointed out by Montagnoli and de Vries (2010), the trading volume on most EU-ETS platforms was thin in Phase I. Significant improvements in liquidity right from the start of Phase II can therefore be readily evidenced on the EEX than on the vastly more liquid ECX. The low trading volumes on the EEX provide a significant advantage for capturing liquidity effects with the analytical methods used (see Hedge and McDermott 2003). Furthermore, among the largely illiquid platforms, it is the only one with accessible microstructure data variables required for this study.

Extant literature on the liquidity of the EU-ETS remains scarce, few published papers, such as Frino et al. (2010), have a direct focus on market liquidity. Frino et al. (2010) find that liquidity improves over time for the EU-ETS. However, their results are not focused on analysing the liquidity effects of the onset of trading in Phase II and other events tested in this book, given that they only focus on quarterly results. This chapter provides daily review of volume and liquidity changes over shorter windows and with respect to the evolution of transaction volumes. A further unpublished manuscript by Benz and

Hengelbrock (2009) (earlier reviewed in the Introduction section) examines liquidity in Phase I of the scheme. These papers are not directly relevant to this study, given that its focus is on the liquidity changes between Phase I and Phase II of the EU-ETS and other events not previously examined for liquidity effects. Other papers related to the study in this chapter are discussed in this book's introduction.

Firstly, abnormal returns over 181 days, straddling the transition to Phase II, are examined using the market model established by Brown and Warner (1985). This means the analysis is concentrated on 90 days on either side of the first day of trading in Phase II. Secondly, volume changes following the trading of the December 2008 futures contract in Phase II are investigated and thirdly, ratios of bid-ask spread liquidity proxies are applied to determine liquidity changes. The basic motivation for applying the bid-ask spread ratios-based methodology is that bid-ask spreads are adequate measures of event impacts on the liquidity of financial instruments. Dennis and Strickland (2003), Pham et al. (2003), Cao et al. (2004), Schrand and Verrecchia (2005) and Lesmond et al. (2008), all employ similar approaches. Liquidity measures play an important role in finance and hence have attracted tests of robustness. Goyenko et al. (2009) in an important paper examine the most commonly used measures of liquidity in the finance literature. When measured against specific benchmarks, liquidity measures based on spreads usually win the horserace on robustness. This chapter implements ratio of the quoted and relative bid-ask spread measures. However, Lee and Ready (1991) indicate that the relative bid-ask spread may be an inaccurate measure of liquidity if many trades occur with the bid and ask prices. As this study uses only daily transactions data from the EEX, one cannot ascertain the volume of trades within the spreads; hence, for robustness, the effective bid-ask spread measure (Hedge and McDermott 2003; Heflin and Shaw 2000) is also implemented.

The results obtained in this chapter suggest that there have been significant liquidity improvements since the transition to Phase II under the new regime of rules. The improvements are not limited to the start of Phase II alone; they are maintained over 90 trading days after the transition. The study also finds significant improvements in trading volumes over the same period. The findings, when viewed in tandem with the European Commission's reports on net short emissions as submitted for compliance year 2008, indicate that the new regime of rules and tighter caps introduced for the Phase II trading period are correlated with improvement of market quality on the hitherto highly illiquid EEX. Also, in order to eliminate the possibility that the analysis captures calendar effects, the study investigates also liquidity effects for the start of trading in 2007 and 2009 with no significant changes recorded for the short term. The introduction of a new platform security-related policy ((EC) No 994/2008 of 8 October 2008) mid-phase seems to have had the opposite effect, however. Results obtained suggest that EC Regulation (EC) No 994/2008 of 8 October 2008 is linked to a significant loss of liquidity in the December 2009 EUA futures contract since there is some evidence of declining short-term liquidity around the event date. For the release of compliance results for years 2007 and 2008, results suggest that the announcements are associated with short and long-term liquidity improvements. None of the events tested show significant positive abnormal returns for any of the EEX EUA futures contracts.

The remainder of this chapter is arranged as follows: Sect. 5.2 discusses the trading environment on the EEX. Section 5.3 reports on the data, Sect. 5.4 discusses the methods and results of the analysis and Sect. 5.5 concludes the chapter.

5.2 The Trading Environment on the EEX

The EEX currently offers trading in emission allowances spot and derivatives as part of their secondary market trading operations. As with the ECX in London, the most traded secondary market instrument at the EEX are the annual December expiry futures contracts. The underlying for each futures contract on the EEX is 1000 EUA (1000tCO$_2$) and the settlement form is delivery versus payment. There are currently two delivery dates in December: early and mid-December, trading has exclusively been on the mid-December contracts. Electronic trading continuously takes place between 0900 and 1700 hours CET on trading days. Prices are quoted in Euros with a minimum tick of €0.01 per tCO$_2$ by market makers. Until recently, the exchange had a single market maker in RWE Supply and Trading GmbH. There are now two market makers each for emissions spot and futures; these are Axpo Trading AG and Belektron d.o.o. for emissions spot and Vattenfall Energy Trading Netherlands N.V.

and Belektron d.o.o. for emissions futures. Although it is a market leader in emissions spot auction volume, electronic trading is perhaps the best option for the EEX in futures trading in view of its comparatively low daily trading averages. In more conventional markets, the introduction of electronic trading with market makers has been found to have greatly improved liquidity (see Barclay et al. 1999; Domowitz 2002).

All transactional settlements and allowances booking within the bounds of the national register (German Emissions Trading Authority) on the exchange are guaranteed by a clearinghouse, the European Commodity Clearing AG (ECC). The EEX offers trading in EUA spot (at present, 2013–2020), EUA futures, EUA options and EUA spreads. It also offers trading in EU Aviation Allowances (EUAA) futures, EUAA spreads, CER spot, CER futures and CER spreads. EUAs, EUAAs and CERs can be submitted for registration on the platform following OTC trading. The platform also offers daily auctions for EUAs and EUAAs in its primary market, alongside continuous trading on the secondary market; the brokered prices on these auctions are published daily as the EEX Carbon Index (Carbix®). The auctions are mainly at the behest of the German Federal Ministry of Environment and 24 other EU countries; they represent the compulsory auctionable NAP allocations for EU installations in those 25 countries.

Trading volume on the EEX has since 2010 advanced rather rapidly. This was first due to the decision of Europe's largest CO_2 emitter, RWE AG to shift some of its carbon trade from the ECX to the EEX (Carr 2010), then the exit of other competing platforms such as Bluenext in France.

5.3 Data

5.3.1 Sample Selection

Firstly, calendar time is converted to event time (see Beneish and Gardner 1995; Gregoriou and Ioannidis 2006). The day2 January 2008, which is officially the first day of trading in Phase II, is defined as event day 0 for the investigation of liquidity effects after the onset of trading in Phase II. The Dec-2008 EUA futures contract that has been trading on the EEX since during the first phase is then selected for this analysis because it satisfies the following conditions and also accounts for more than 85% of trading volume during the period: (1) The contract has historical data for a period of 90 trading days before and after the event; (2) The contract is the most actively traded contract on the exchange 90 trading days before and after the event. A scarcity of transactions on the EEX is observed such that at any one period only one or two contracts are sufficiently traded for rigorous statistical analysis; hence, events are examined using the most actively traded contract during the period which the event occurred.

The study is extended to include three other events of policy interest (see Table 5.1).[1] Conditions (1) and (2) above are applied in the selection of EUA futures contract to be examined for the liquidity effects of the other investigated events. The events chosen must satisfy the following condition only: (1) No other event/announcement of relevance to the market occurs within 90 days before and after the chosen event in order to avoid the problem of confounding effects.

Daily data on trading volume for all futures contracts traded on the EEX over the time period of 4 October 2005 to 30 December 2010 is obtained through manual population from EEX's website. The data period extends beyond this chapter's immediate enquiry for the purpose of providing a descriptive view of market variables over an extended period (see Table 5.2). Daily best bid and ask prices are obtained along with daily trading variables (including volume and open interest) and the last transaction price and time for each day. The only intraday variable is the last transaction price for each day; this is required in estimating the effective bid-ask spread at the close.

5.3.2 Sample Description

In Panels A and B of Table 5.2, summary statistics for the market liquidity proxies and trading activity measure of daily traded volumes are presented. Panel A is for Phase I and Panel B is for Phase II. The evidence points to higher intertemporal fluctuations in daily volume than for the liquidity proxies; this is implied by the higher coefficients of variation for

Table 5.1 Table of events and corresponding dates

Date	Event tag	Event description	Contract used in testing impact of event
02/01/2008	Event 1	First day of trading EUA futures contracts in Phase II	EEX EUA Futures Dec-2008
08/05/2008	Event 2	Release of verified emissions retirement data for compliance year 2007	EEX EUA Futures Dec-2008
08/10/2008	Event 3	'Commission Regulation (EC) No 994/2008 of 8 October 2008 (Amendment to the Commission regulation (EC) No 2216/2004 of 21 December 2004) for a standardised and secured system of registries pursuant to Directive 2003/87/EC of the European Parliament and of the Council and Decision No 280/2004/EC of the European Parliament and of the Council'	EEX EUA Futures Dec-2009
11/05/2009	Event 4	Release of verified emissions retirement data for compliance year 2008	EEX EUA Futures Dec-2009

The table lists the events that are examined for their liquidity impacts on EEX EUA futures contracts. The contracts used for each event are listed in the fourth column of the table

Table 5.2 Descriptive statistics

Panel A: Phase I

	Quoted spread	Relative spread	Effective spread	Daily volume (contracts)
Mean	46.31%	3.62%	32.87%	46.26
Standard deviation	0.222	0.031	0.285	81.803
Coefficient of variation	0.518	1.171	1.139	8.180
Median	43.00%	2.62%	25.00%	10.00

Panel B: Phase II

	Quoted spread	Relative spread	Effective spread	Daily volume (contracts)
Mean	13.48%	0.77%	19.74%	269.560
Standard deviation	0.075	0.003	0.195	359.773
Coefficient of variation	0.627	0.448	1.496	2.636
Median	12.00%	0.74%	13.00%	136.50

Panel C: Liquidity proxies

Quarter	Quoted spread (%) mean (median)	Relative spread (%) mean (median)	Effective spread (%) mean (median)	Depth	€Depth	CompositeLiq (%)
4th 2005	64.00 (60.00)	2.92 (2.63)	42.60 (40.00)	6.45	142.94	2.04
1st 2006	59.00 (60.00)	2.20 (2.11)	29.83 (25.00)	14.08	369.29	0.60
2nd 2006	68.00 (62.00)	3.45 (3.37)	62.19 (45.00)	9.92	205.71	1.68
3rd 2006	44.00 (45.00)	2.55 (2.59)	22.37 (15.00)	7.18	125.03	2.04
4th 2006	51.00 (43.00)	3.58 (3.04)	27.10 (10.00)	12.52	179.36	1.99
1st 2007	36.00 (37.00)	2.14 (2.62)	20.83 (15.00)	10.03	113.90	1.88
2nd 2007	35.50 (33.00)	2.75 (1.95)	28.15 (15.00)	47.83	997.13	0.28
3rd 2007	36.00 (36.70)	8.10 (7.70)	34.35 (30.00)	81.91	1671.68	0.48
4th 2007	23.00 (21.00)	3.30 (1.30)	27.66 (30.00)	49.77	1117.78	0.30
1st 2008	17.00 (12.70)	0.76 (0.57)	15.77 (12.00)	111.76	2423.24	0.03
2nd 2008	17.00 (16.80)	0.62 (0.63)	16.50 (11.00)	86.55	2260.25	0.03

(continued)

Table 5.2 (continued)

Panel C: Liquidity proxies

Quarter	Quoted spread (%) mean (median)	Relative spread (%) mean (median)	Effective spread (%) mean (median)	Depth	€Depth	CompositeLiq (%)
3rd 2008	21.00 (21.30)	0.81 (0.84)	23.00 (17.50)	179.03	4397.99	0.02
4th 2008	23.00 (22.00)	0.99 (1.02)	24.12 (19.00)	404.80	7795.26	0.01
1st 2009	14.00 (11.50)	1.10 (0.99)	23.42 (15.00)	72.33	837.59	0.13
2nd 2009	14.60 (14.30)	0.98 (0.94)	28.67 (16.00)	67.15	939.52	0.10
3rd 2009	13.00 (14.00)	0.88 (0.88)	16.38 (8.50)	43.20	622.33	0.14
4th 2009	12.80 (12.30)	0.88 (0.86)	20.64 (13.00)	57.30	797.81	0.11
1st 2010	10.50 (10.30)	0.77 (0.76)	21.57 (15.00)	133.83	1761.77	0.04
2nd 2010	10.20 (10.50)	0.65 (0.66)	24.35 (17.70)	194.13	2971.22	0.02
3rd 2010	7.10 (6.30)	0.47 (0.41)	14.58 (10.50)	407.62	6082.40	0.01
4th 2010	6.50 (5.30)	0.43 (0.35)	11.03 (8.00)	295.21	4466.57	0.01

Panels A and B show the descriptive statistics for daily bid-ask spread values for EEX EUA futures contracts trading on the EEX platform between 4 October 2005 and 30 December 2010. Panels A and B are for Phase I and Phase II, respectively. Panel C shows daily liquidity measures per quarter using data over the same period. The quoted bid-ask spread is the difference between the daily best ask and bid prices, the relative bid-ask spread is the daily ask price minus the best bid price divided by the daily best mid-quote, effective bid-ask spread is twice the absolute value of the prevailing transaction price minus the daily best mid-quote. Coefficient of variation is the ratio of standard deviation to the corresponding median value. Depth (€Depth) is computed as the difference between open interest (euro value of open interest scaled by 1000) at t minus $t-1$. CompositeLiq is the value of %relative spread divided by €Depth. All values are computed by obtaining the averages for each individual contract trading during the period and then cross-sectionally aggregating across the range of contracts. Each EUA futures contract has an underlying of 1000 tonnes of CO_2. Excluding discarded non-trading dates, the panels contain data for 1280 trading days the volume variable. This may be due to the fact that bid and ask prices are essentially discrete variables; this property helps in diminishing the potential for volatility through price clustering (Chordia et al. 2001). Furthermore, in Panel A, there is a substantial level of variation between the relative and effective bid-ask spread values. This implies that a vast proportion of carbon trading occurs within the ask and bid quotes provided by the market maker.

It is also of interest to note that the high levels of bid-ask spreads and the low average trading volume of the carbon permits in Panel A suggest that the emission permits are less liquid in Phase I of the EU-ETS, in comparison with Phase II. This is not surprising in view of the lack of trading volumes linked with the overallocation of carbon permits. To provide an expansive view of liquidity changes in the EU-ETS, quarterly measures of six liquidity proxies are provided in Panel C of Table 5.2 as part of the descriptive analyses. The Panel shows that liquidity has advanced considerably over Phase II.

5.4 Results and Discussion

5.4.1 Abnormal Return of Events: Dec-2008 and Dec-2009 Contracts

The event study methodology, with the market model, as outlined in Campbell et al. (1997) is employed in estimating abnormal returns for the EEX EUA futures contracts with December 2008 and 2009 expiries, that is, the Dec-2008 and Dec-2009 futures traded on the EEX. The market model has been used by several studies (e.g. see Brown and Warner 1985; Denis et al. 2003; Gregoriou and Ioannidis 2006; Hedge and McDermott 2003) within an event study approach.

The market model assumes a linear correlation between the return of a given security with a value-weighted index. The model is given for any asset i as:

$$R_{it} = \alpha_i + \beta_i R_{mt} + \varepsilon_{it} \qquad (5.1)$$

$$E[\varepsilon_{it}] = 0 \quad \text{Var}[\varepsilon_{it}] = \sigma_{\varepsilon i}^2,$$

where R_{it} and R_{mt} are the time t returns on asset i and the market portfolio, respectively. ε_{it} is a zero mean disturbance term, while α_i, β_i and $\sigma_{\varepsilon i}^2$ are parameters of the model. Abnormal return is obtained from the market model as follows:

$$\varepsilon_{it}^* = R_{it} - E[R_{it} | X_t] \tag{5.2}$$

R_{it}, $E[R_{it}]$ and ε_{it}^* correspond to the actual, normal and abnormal returns, respectively. X_t is the conditioning information corresponding to the market return (Campbell et al. 1997). The LEBA carbon index is used in estimating the model parameters over 90 days prior to the events. The index is computed for every trading day in the sample employing value-weighted average of all carbon trades executed by LEBA firms. Abnormal return for each trading day in the event windows are obtained after OLS estimation with Newey and West (1987) HAC. These are then aggregated through time to obtain the cumulative abnormal return (CAR) for each window investigated.

The average abnormal returns (AAR) for each event window are reported in Table 5.3. The results in the second column of the table suggest that the onset of trading in Phase II is not associated with significant abnormal returns for the Dec-2008 although the abnormal returns are positive in the short term. In the long term, the abnormal returns are negative and also not significant. Table 5.3 also reports results for Events 2, 3 and 4. The results indicate that none of the events are related to significant abnormal returns for the tested December maturity contracts; however, again, a number of the abnormal return estimates are positive. The predominantly positive values although not significant may be construed as a suggestion of price appreciation due to the events (especially with Event 1). The negative long-term AAR values, however, indicate that if this were accurate, the price improvements in any case are not permanent. According to Campbell et al. (1997), the large R^2 values obtained in the market model estimation indicate corresponding variance reduction thereby leading to gain in model specified. The R^2 values are 0.81, 0.58, 0.62 and 0.72 for Events 1, 2, 3 and 4, respectively.

Table 5.3 Average abnormal returns

Event window	Event 1 AAR (%)	Event 2 AAR (%)	Event 3 AAR (%)	Event 4 AAR (%)
[−1, +1]	0.48	−0.20	−0.23	−0.21
	1.02	−0.40	−0.29	−0.12
[−2, +2]	0.36	−0.16	0.49	0.40
	1.23	−0.44	0.78	0.36
[−3, +3]	0.29	−0.25	0.17	0.40
	1.27	−0.89	0.36	0.51
[−4, +4]	0.08	−0.08	0.02	−0.09
	0.36	−0.28	0.06	−0.12
[−5, +5]	0.00	0.10	0.06	−0.03
	0.001	0.39	0.18	−0.04
[0, +10]	−0.25	0.10	0.02	0.36
	−0.98	0.43	0.03	0.45
[0, +20]	−0.31	0.14	−0.12	0.03
	−0.71	0.78	−0.28	0.06
[0, +30]	−0.10	0.06	−0.21	0.05
	−0.3	0.43	−0.66	0.15
[0, +60]	−0.04	−0.07	−0.06	0.03
	−0.19	−0.40	−0.26	0.13
[0, +90]	−0.02	−0.03	−0.07	0.03
	−0.12	−0.21	−0.34	0.22
Adjusted R^2	0.81	0.58	0.62	0.72
BG-LM	0.00	0.08	0.06	0.11
BPG	0.19	0.12	0.17	0.54

Average abnormal returns (AAR) using the market model (estimated using OLS with Newey and West 1937 HAC) are computed for an event study aimed at determining excess returns around the events' days. The estimation window for estimating the model parameters is 90 days before and after the events (−90, +90). AAR is then tested for being significantly different from zero by with a regular t-statistic. t-Statistics are reported underneath the AAR values in each corresponding event window and event boxes. Event 1 is the transition of trading from Phase I to Phase II. Event 2 and Event 4 are dates for the release of emissions verification results for compliance years 2007 and 2008, respectively. Event 3 is the date for the adoption of European Commission Regulation (EC) No 994/2008 of 8 October 2008. Event 1 and Event 2 are investigated using the December 2008 contract, and Event 3 and Event 4, the December 2009 contract. BG-LM and BPG are the p-values for the Breusch–Godfrey serial correlation and the Breusch–Pagan–Godfrey (heteroscedasticity) LM test statistics, respectively. One EUA Futures contract has an underlying of 1000 EUAs

5.4.2 Impact of Events on Trading Volumes: Dec-2008 and Dec-2009 Contracts

5.4.2.1 Short-Term Impact of Events

The presence of abnormal trading volume in the event period is investigated by employing the following dummy time series regression model:

$$\text{Volume}_t = \alpha + \sum_{-5}^{+5} D_i \beta_i + \varepsilon_i \text{ for } t = -90, +5, \quad (5.3)$$

where Volume_i is the logarithmic transformation of trading volume for the futures contract at day t. α is a constant, Di are dummy variables for each trading day in the event window $[-5, +5]$. The coefficients of the 11 dummy variables, β_i capture the abnormal trading volume over the event window, $[-5, +5]$, ε_i is a random disturbance term with a mean of zero and a variance σ^2. α and β_i are parameters to be estimated. Equation (5.3) is estimated by OLS with White's (1980) heteroscedastic consistent covariance matrix, and for robustness, also Newey and West (1987) HAC, the statistical inference levels obtained are significantly the same.

The results for the time series regressions are reported in Table 5.4. The positive and significant sign of the 11 dummy variables suggests an improvement in trading volumes of carbon permits being associated with the start of trading in Phase II. Following the introduction of Phase II, volume continues to increase on a significant basis throughout the post announcement period covered by our model. The association is underscored by the fact that 2 and 3 January 2008 (the first and second days of trading in Phase II) have two of the three largest coefficient estimates (t-statistics) in the 11-day period examined with 2.27 (10.82) and 2.28 (10.87), respectively. Thus, the estimates are statistically significant at the 1% level. After 3 January 2008, the abnormal volume decreases from the peak values but remains positive and significant at the 5% level.

The results for Events 2, 3 and 4 are presented in the last three columns of Table 5.4. For Events 2 and 4, there are significant positive estimates reported once the emissions verification results were released.

The Liquidity Effects of Trading Carbon Financial Instruments

Table 5.4 Short-term trading volume changes around events

Parameter	Event 1 estimates	Event 2 estimates	Event 3 estimates	Event 4 estimates
A	3.00	4.90	2.71	3.99
	14.33***	51.54***	11.88**	23.57***
β_{-5}	1.55	0.28	−2.71	−1.59
	7.39***	2.96***	−11.88**	−9.39***
β_{-4}	1.70	−0.36	1.20	−3.99
	8.09***	−3.77***	5.27**	−23.57***
β_{-3}	1.09	0.52	1.30	0.57
	5.20**	5.42***	5.69**	3.35**
β_{-2}	3.00	0.04	2.64	0.19
	14.33**	0.43	11.57**	1.11
β_{-1}	0.91	−0.74	4.31	0.83
	4.33***	−7.82***	18.91**	4.93***
β_0	2.27	0.18	3.76	0.52
	10.82***	1.93*	16.49**	3.10***
β_{+1}	2.28	0.46	−2.71	1.68
	10.87***	4.85***	−11.87**	9.96***
β_{+2}	1.90	0.46	−2.71	0.51
	9.06***	4.85***	−11.88**	3.03***
β_{+3}	1.42	0.71	−0.41	0.19
	6.75***	7.50***	−1.78*	1.11
β_{+4}	1.38	0.19	1.51	0.99
	6.57***	2.02**	6.62**	5.86***
β_{+5}	1.94	0.56	2.98	1.08
	9.24***	5.87***	13.06**	6.36***
R^2	0.12	0.04	0.18	0.12
NORM (1)	0.17	0.10	0.17	0.11
ADF	0.00	0.00	0.00	0.00

The following time series regression model is estimated (using both White's (1980) heteroscedastic consistent covariance matrix and Newey and West (1987) HAC alternatively) to examine trading volume changes around the events' days on specific EEX EUA futures contracts:

$$\text{Volume}_t = \alpha - \sum_{-5}^{+5} D_i \beta_i + \varepsilon_i \text{ for } t = -90, +5$$

where Volume_t corresponds to the log of the traded volume for day t. α captures the trading volume variations over the 96 day estimation period and D_i are dummy variables representing each day in the investigated event window (−5, +5). The coefficients of all 11 dummy variables encapsulate the variations in trading volume over the event window, one for each day in the event window (−5, +5). ε_i is a residual term with $E[\varepsilon_i] = 0$ and $\text{Var}[\varepsilon_i] = \sigma^2$. α and β_i are parameters

(continued)

Table 5.4 (continued)

for estimation. The estimates are reported in the boxes with corresponding t-statistics underneath. Event 1 is the transition of trading from Phase I to Phase II. Event 2 and Event 4 are dates for the release of emissions verification results for compliance years 2007 and 2008, respectively. Event 3 is the date for the adoption of European Commission regulation (EC) No 994/2008 of 8 October 2008. Event 1 and Event 2 were investigated using the December 2008 contract, and Event 3 and Event 4, the December 2009 contract. NORM (1) and ADF are the p-values for the Jarque–Bera normality test and the Augmented Dickey–Fuller test statistics, respectively. The lag length for ADF is selected on the basis of Schwarz information criterion (Schwarz 1978)

*, ** and *** correspond to statistical significance at 10, 5 and 1% levels, respectively

This indicates relative upsurge in trading volumes. For Event 3, three of the post-event estimates are negative with two positive. All estimates are statistically significant; thus, the result is not conclusive with respect to association of the event with trading volumes. The liquidity effects are thus examined in Sect. 5.4.2.2. The regression for Eq. (5.3) also passes the normality test for all four events, implying that the abnormal volume empirical estimates are not as a result of the presence of outliers in the data. Also the Augmented Dickey–Fuller test (see Dickey and Fuller 1979) show that the null of presence of unit root is strongly rejected for the log (volume) variable in each case, hence, the results are not spurious. The reported p-values are (MacKinnon 1996) one-sided p-values.

5.4.2.2 Long-Term Impact of Events

In order to analyse changes in the long-term trading volume of carbon permits preceding Phase II EU-ETS trading and other events, post/pre ratio of long-term trading volume in the post-event period [0, +90] to the long-term volume in the pre-event period [0, −90] is constructed. If there is sustained increase in trading volume (over 90 trading days) when the Dec-2008 contracts are traded in Phase II as against when traded in Phase I, this may lead to increasing economies of scale for the market maker, resulting in lower bid-ask spreads and higher market liquidity (Copeland and Galai 1983).

Table 5.5 Long-term trading volume around events

Variable	Long-term trading volume			
	Event 1	Event 2	Event 3	Event 4
Mean (post/pre)	5.28	1.50	4.72	2.66
Median (post/pre)	5.21	1.34	4.62	2.59
t-statistic	13.05***	2.20**	3.76***	2.62**

The sample consists of EEX EUA futures contracts traded before and after the introduction of Phase II of the EU-ETS. Changes in long-term trading volume are defined as trading volume in futures contracts in the post-event period [0, +90] divided by the trading volume in futures contracts in the pre-event period [0, −90]. The t-statistic is constructed to test the null hypothesis that the trading volume is unchanged in the pre-event period as compared with the post-event period
** and *** correspond to statistical significance at 5% and 1% levels respectively

The results from analysis of long-term changes in trading volume are reported in Table 5.5. The mean (median) post/pre ratio of trading volume is 5.28 (5.21) for Event 1, this is with a corresponding t-statistic of 13.05. This is another evidence of the association of commencement of trading in Phase II with improvements in trading volumes. For Events 2 and 4, which are the announcement dates for verified emissions results, statistically significant post/pre ratios are also observed, further underscoring the positive association of the release of the events with trading volume improvements. The respective mean (median) values for post/pre ratio for Events 2 and 4 are 1.50 (1.34) and 2.66 (2.59).

Another interesting part of the result is the recording of large and statistically significant post/pre ratios for Event 3. In this chapter, we examine this in subsequent sections to see whether the observed long-term improvement is accompanied by long-term improvements in liquidity. The estimates in Tables 5.4 and 5.5 provide very little indication of what to expect since liquidity and trading volumes are weakly related, in fact, the innovations of both variables can be conditionally uncorrelated (see Johnson 2008). In Sect. 5.4.5, the liquidity effects are examined; however, some discussions on measuring financial markets liquidity are first presented in Sects. 5.4.3 and 5.4.4.

5.4.3 Financial Markets Liquidity

In the introduction to this book, we discuss the common basis that liquidity as a concept shares with the other key microstructure function of financial markets, that is, price discovery. In subsequent sections, we focus on how to proxy or quantify liquidity in a market such as the EEX platform. According to Sarr and Lybek (2002), liquid securities/markets usually manifest some or all of the following five features: tightness, immediacy, depth, breadth and resiliency. Tightness refers to the difference between what the buyer and seller of an instrument believe it is worth. Thus, quote-driven markets, the narrower the spread between the ask and bid prices, the lower the transaction costs and hence the higher the liquidity in the market for the relevant instrument. Immediacy refers to the speed of executing an order with little or no price impact. Depth (orders at various bid and ask prices) and breadth (range of counterparties) are both reflected by the availability of orders and large volumes of securities ready for transactions as required. There will always be a degree of disproportional trading with respect to generating buy and sell orders especially on short-run basis (Chordia and Subrahmanyam 2004); resiliency refers to the market's ability to generate new orders in order to correct such short-run inconsistencies. These five features correspond to Kyle's (1985) three features of liquidity, which include tightness, depth and resilience.

The above description quite captures the qualitative concept of liquidity but the quantitative assessment is still unsettled in the academic literature (e.g. see Baker 1996; Hallin et al. 2011; Sarr and Lybek 2002). And as the role of liquidity in empirical finance has matured to the extent of critically affecting decisions on asset pricing, general market efficiency and corporate finance, so has the interest it generates among practitioners and academics alike grown (see Goyenko et al. 2009). Measuring financial markets liquidity is contingent on the possibility of substitutability of asset classes traded on that market and the extent of liquidity of the assets. The availability of a range of issuers for individual assets can create a difficulty leading to fragmentation in the market, thereby engendering a loss of fluidity. This is not the only obstacle to measuring market liquidity based on the liquidity measures of component assets; similar asset classes having the same issuer may also have observable peculiarities under differing conditions. Examples of these potential differences include varying maturities in the case of futures and options. These represent a fundamental problem involved in blending individual measures of liquidity for different assets together in order to arrive at an acceptable and unbiased measure of market-wide liquidity. This has led to a relatively limited number of studies being carried out on market-wide liquidity over the years (see Chordia et al. 2001). However, some of the problems are solvable; for example, if market liquidity is treated as a dynamic factor, time dependence and commonness can be controlled for simultaneously (see Hallin et al. 2011).

5.4.4 Measures of Liquidity

Three key groupings of liquidity measures in finance literature (see for discussions Chordia et al. 2001; Grossman and Miller 1988; Hallin et al. 2011; Hasbrouck and Schwartz 1988; Sarr and Lybek 2002) are discussed in this section.

5.4.4.1 Volume-Based Measures

Although a higher volume of transactions is not necessarily correlated with improvements in market liquidity (see Johnson 2008). Volume-based measures are commonly used as proxies for market liquidity. For example, in Panel C of Table 5.1, two depth measures are used as liquidity proxies. In addition, turnover is employed as a proxy for liquidity the regression analysis presented in Chap. 4. This is tenable since market transactions executed in significant numbers provide vital order flow information to market participants. For market makers in a quote-driven market, this allows for the evaluation of the degree of correctness of their quotes. Changes in quotes, prompted by observing the order flow, ultimately lead to a re-alignment of the order flow in order to reflect an efficient price signal. This implies that the market is resilient. This process is continuous, consistently supplying the needed information to the participants in order to effect an efficient price discovery.

Markets lacking in both depth and breadth cannot effectively provide the above described vital stream of information to market participants, and this has huge implications for market efficiency. Frequent trade interruptions and general lack of price continuity evident on most emissions trading platforms in Phase I of the EU-ETS and to some extent in Phase II is informed by the absence of market depth and breadth. Issues such as this breed unpredictability in markets, especially when it comes to price signalling. Uncertainty about price signals can however be overcome by looking to similar markets/platforms with more depth and wider breadth as is the case in the EU-ETS, and this is where substitutability of assets plays a key role. The ICE ECX platform was responsible for about 90% of exchange-based transactions in the EU-ETS during most of the second phase of the scheme (see Daskalakis et al. 2011) and thus led the price discovery process (see Mizrach and Otsubo 2014; Benz and Hengelbrock 2009) and continues to lead the price discovery process in the third phase of the EU-ETS. Therefore, thin trading volumes on less liquid EU-ETS platforms need not inhibit price efficiency as price signals from the ECX can be employed on other platforms for trading.

Furthermore, an efficient price discovery process may not be solely dependent on liquidity. In a case where buyers and sellers as well as their liquidity requirements are easily discernible, sufficient trading can occur without an endogenous pull of the order flow. This appears to be the case for the EU-ETS. The market is mandatory for some 12,000 participants and there are others who trade for other purposes, such as risk hedging and for liquidity. These participants who must trade for compliance purposes in the market are readily identifiable; hence, the thriving of OTC trades at the start of the EU-ETS in Phase I, even when exchange-based trading was virtually inexistent.

5.4.4.2 Price Movement and Market Resilience Measures

Price movements in a market can result from liquidity fluctuations in the market; the fluctuations may also be exogenously triggered. This category of measures aims to distinguish between liquidity influenced price movements and those that occur as a result of other unaccounted for factors.

These measures relate to price discovery (efficiency) and resilience. Most price-based proxies of liquidity harness a fundamental view of price movements in liquid and illiquid markets. There is a permanent element to price changes in liquid markets that is absent in illiquid ones. However, there will be some form of transient changes to prices even in a liquid market due to exogenous shocks. Shock-induced jumps affect price discovery, nevertheless, the continuity in price formation should still be sustained (see Sarr and Lybek 2002). Hence, when a lasting price shift occurs, the effect should be such that the temporary elements of the price change will be very little in comparison to an illiquid markets. The market-efficiency coefficient (MEC) of Hasbrouck and Schwartz (1988) exploits this element of price change. It describes a coefficient differentiating transitory and permanent price shifts. The coefficient is a process expressed as:

$$\text{MEC} = \text{Var}(R_t) / [T \times \text{Var}(r_t)], \quad (5.4)$$

where $\text{Var}(R_t)$ is the variance of the log of long-period returns, $\text{Var}(r_t)$ is the variance of short-period returns and T corresponds to the number of short periods in each longer period. The ratio of markets exhibiting higher levels of resilience (liquidity) should be slightly lower than one. In those with low levels of resilience, the ratio will be considerably lower than one. The difference in ratios is accounted for by the variance in short-period price volatility in the different markets. Bernstein (1987) identifies several factors influencing disproportionate short-period price volatilities; these include interventions in the price formation process by market makers, errors in price discovery, which may also be linked to meddling by market makers. Erroneous price formation includes lagged price reaction to pertinent announcements/events bringing about positively correlated price shifts incrementally. This diminishes short-period volatility in correspondence to long-period volatility, ultimately leading to MEC rising above one.

The MEC is basically a variance ratio test of the random walk. The variance of long-horizon returns is divided by the variance estimated for returns over shorter intervals. For a market in harmony with the random walk process, the variance of returns measured over longer horizons is equal to the sum of variances of shorter horizon returns as long as the summation of the shorter horizons is equal to that of the longer horizon. Thus, variance of the longer horizon returns is η times the variance of returns measured over shorter horizons, if η is the number of short-horizon periods in the longer horizon. According to Grossman and Miller (1988), a divergence from the random walk can be induced by inventory-related issues due to return serial correlation; however, in a largely efficient market, arbitrage opportunities created by this deviation will lure participants into providing required trading volumes. Hence, the divergence from random walk will be very much temporary even if market makers cannot absorb orders.

In building a case for the use of price-based measures as proxies for liquidity, this book has sought to tie in price continuity with market resilience. Market resilience and price continuity are not exact alternates. Participants in the market trade either for liquidity purposes or to take advantage of private information. Thus, the positions they take are reflected in the order flow in accordance with their motives for trading. Irrespective of these positions, they all respond to shifts in fundamentals, although markets do become skewed without being prompted by changes to fundamentals, as noted by Grossman and Miller (1988). If a market becomes dislocated from fundamentals, a price shift will occur. Expectedly, a resilient market will promptly generate the required orders (mainly arbitrage market orders) to correct the imbalance in orders and the corresponding price shift. The price is therefore prevented from further progressing on its current course by the influx of new orders aimed at taking advantage of an arbitrage opportunity, but ultimately correcting the order imbalance (see Chordia et al. 2002; Chordia et al. 2008). This sequence underscores the connection between market resilience and price continuity.

5.4.4.3 Transaction Cost Measures

Analysis of transactions prices provides a means of distinguishing several cost factors, such as costs associated with order processing, information asymmetry and inventory. In quote-driven markets, the market maker quotes provide a basis for measuring the transactional (bid-ask) spread. Within the spread lie certain cost components, which reflect the quality of the trading process. These cost components include asymmetric information/adverse selection costs, inventory costs and order processing costs. Campbell et al. (1997) identify order processing, inventory and adverse selection costs as the three basic economic information sources in market microstructure models.

Order processing costs are associated with the direct price of executing transactions in a given market. Inventory costs are those that come with having to hold the financial securities, usually costs associated with record keeping in the case of most financial securities. In the EU-ETS and similar schemes, the basic inventory cost incurred will be the cost of operating and setting up recording devices to keep track of transactions. Adverse selection costs, however, refer to the costs incurred by trading with a more informed counter-party. From the perspective of the market maker, some investors/traders may be sophisticated enough to generate private information and then use such to place orders in the market. Trading with such a counter-party implies a loss of income due to information asymmetry in the market as induced by the existence of the provide information to a counter-party. This loss of income is seen as the cost of being adversely selected by an informed trader. Market makers are required to provide liquidity in a quote-driven market and hence must trade with all participants when the need arises. With no ability to discern if a participant is better informed than them or otherwise, it becomes appropriate that the market maker offer spreads that account for the risk of being adversely selected, thus adverse selection costs are incorporated.

Based on the varying properties of these costs, they possess distinctive statistical characteristics. The distinctiveness of their properties spurred the growth of the asymmetric information and inventory costs market microstructure literature streams in the 1970s and 1980s. This development provided unambiguous projections on computing the bid-ask spread (see as examples Glosten 1987; Glosten and Milgrom 1985; Ho and Stoll 1981, 1983; Roll 1984). Advancements were also recorded in deriving adequate estimators for spread components based on the existing theories (see as examples Foster and Viswanathan 1993; Glosten and Harris 1988; Hasbrouck 1991a, b; Stoll 1989).

Demand for transactions is usually adversely affected by large trading costs; indeed a characteristic of most liquid markets is low transaction costs. Reasonably low transaction costs can engender diversification, a vital condition for the sustenance of market liquidity in a scheme like the EU-ETS. High costs of trading in a market can however result in fractured markets since a great deal of trades will be executed inside the spreads rather than at about the equilibrium price (all available information considering). This can consequently result in a market lacking depth (low liquidity). Moreover, wide spreads which signify high transaction costs or presence of informed trades can force participants to seek trading opportunities on parallel platforms thus leading to thin trading. A low level of participation in markets means that the market is thin and therefore lacking breadth. Resilience in the market can also be threatened since the flow of orders will be reduced thereby reducing the capacity of the market to rectify market imbalances in the order flow.

The bid-ask spread is a cost measure proxy for liquidity in empirical finance. There are several variants of the bid-ask spread in the market microstructure literature, but perhaps the most popular is the quoted bid-ask spread. This is a measure of the difference between the lowest ask price and the highest bid price during a specific period t for a security i. This measure captures the economic significance of transactions to the market maker. Sometimes, valid market maker-type quotes can arise in the form of a regular trader issuing limit orders. Other variants of the quoted spread include weighted transaction prices. In a market, where there is more than one market maker that are not compelled to issue the same quotes or even trade at those quotes, distinction should be made between market maker spreads and the market spread; outliers should also be overlooked in computing a single spread for the market.

Needless to state, however, no singular measure can capture all the relevant qualitative dimensions of liquidity, that is, breadth, resilience, depth, immediacy and tightness. Indeed, Hallin et al. (2011:1) identify the task of achieving concurrence on liquidity measures as: '… *double difficulty (which) seriously challenges the objectivity of any final assessment*'. In Sect. 5.4.5, liquidity proxies based on the bid-ask spread are computed in order to observe the liquidity effects of the events investigated in this chapter on the Dec-2008 and Dec-2009 contracts.

5.4.5 Liquidity Improvements: EEX EUA Dec-2008 and Dec-2009 Contracts

Given the short- and long-term changes in the trading volume effects of EEX EEA futures due to the introduction of Phase II of the EU-ETS, as reported in Tables 5.4 and 5.5, market liquidity effects of the events are now tested. In order to analyse the impact of the start of Phase II EU-ETS trading and other events on the short-term liquidity of EEX EUA futures, ratios of the daily average quoted, relative and effective bid-ask spreads over various event windows in the pre and post Phase II trading periods are constructed. The relative bid-ask spread is computed as the ask price minus the bid price divided by the mid-price. However, as pointed out by Lee and Ready (1991) the relative bid-ask spread has two potential shortcomings. Firstly, it overstates the trading costs of securities because it fails to account for the tendency of prices to rise following a purchase and fall following a sale. Secondly, it can be argued that the relative bid-ask spread is an inappropriate measure of instrument liquidity due to the fact that trades frequently occur within the ask and bid prices. Therefore, in order to account for these two shortcomings, the effective bid-ask spread Hedge and McDermott (2003) is also computed. Effective bid-ask spread is defined as twice the absolute value of the difference between the final transaction price in a period and the prevailing mid-price corresponding to that transaction. There is also a potential problem with the use of either the relative or the effective bid-ask spread. The problem is that any event that changes the mid-price, will automatically impact upon the relative and effective bid-ask spreads. Therefore, for completeness, the quoted bid-ask spread defined as the ask price minus the bid price is also computed. The quoted spread is immune to this problem. The quoted bid-ask spread is also constructed because it is a measure that encapsulates the economic significance of trades to the market maker (Branch and Freed 1997).

There is evidence from Table 5.6 that bid-ask spreads are significantly reduced after the introduction of Phase II EU-ETS trading. For example, in the [−5, +5] event window the mean and median quoted bid-ask spread ratios are 0.59 and 0.55, respectively, and are both highly statistically significant at the 1% level. This indicates that trading spreads are

Table 5.6 Short- and long-term liquidity changes on the EEX

Event window	Event 1 Quoted spread ratios	Relative spread ratios	Effective spread ratios	Event 2 Quoted spread ratios	Relative spread ratios	Effective spread ratios	Event 3 Quoted spread ratios	Relative spread ratios	Effective spread ratios	Event 4 Quoted spread ratios	Relative spread ratios	Effective spread ratios
[−1, +1]	0.74 (0.73) −2.25**	0.70 (0.67) −2.82**	0.43 (0.37) −9.23***	0.66 (0.66) −200.0***	0.58 (0.58) −383.17***	0.40 (0.34) −10.52***	1.22 (1.02) 0.58	1.39 (1.14) 0.89	1.22 (0.95) 0.43	0.74 (0.67) −3.52**	0.52 (0.48) −9.13**	0.29 (0.38) −7.51**
[−2, +2]	0.60 (0.55) −3.49**	0.57 (0.53) −3.99**	0.35 (0.37) −10.70***	0.66 (0.66) −201.1***	0.57 (0.58) −333.52***	0.48 (0.51) −8.12***	1.34 (1.02) 1.21	1.54 (1.14) 1.63	1.02 (0.95) 0.07	0.56 (0.67) −3.01**	0.54 (0.48) −7.70***	0.35 (0.38) −9.26***
[−3, +3]	0.59 (0.55) −4.71***	0.56 (0.53) −5.23***	0.35 (0.37) −14.91***	0.58 (0.66) −8.14***	0.51 (0.58) −10.75***	0.48 (0.51) −9.98***	1.23 (1.02) 1.07	1.40 (1.14) 1.58	0.96 (0.95) −0.15	0.78 (0.78) −3.76***	0.57 (0.59) −9.74***	0.41 (0.38) −6.01***
[−4, +4]	0.59 (0.55) −5.07***	0.56 (0.53) −5.55***	0.32 (0.30) −15.92***	0.68 (0.66) −4.01***	0.61 (0.58) −5.42***	0.58 (0.51) −4.30***	1.23 (1.02) 1.29	1.39 (1.14) 1.88*	0.87 (0.60) −0.62	0.80 (0.78) −3.27**	0.59 (0.59) −8.58***	0.55 (0.38) −3.67***
[−5, +5]	0.59 (0.55) −5.34***	0.56 (0.53) −5.82***	0.34 (0.30) −12.92***	0.70 (0.66) −3.84***	0.62 (0.58) −5.26***	0.65 (0.60) −3.48***	1.19 (1.02) 1.25	1.34 (1.14) 1.95*	0.83 (0.60) −0.93	0.90 (0.89) −1.10	0.67 (0.63) −4.78***	0.56 (0.38) −3.48***
[0, +10]	0.40 (0.33) −8.68***	0.37 (0.30) −9.62***	0.45 (0.37) −4.83***	0.78 (0.66) −2.20**	0.68 (0.58) −3.60**	0.51 (0.51) −4.03***	0.94 (0.70) −1.35	1.09 (0.78) 0.43	0.61 (0.23) −2.08**	0.83 (0.78) −3.28***	0.80 (0.59) −1.30	0.80 (0.43) −0.78
[0, +30]	0.40 (0.29) −12.97***	0.41 (0.30) −12.17***	0.56 (0.41) −4.58***	0.81 (0.77) −3.67**	0.68 (0.64) −7.41***	0.74 (0.60) −2.92**	0.96 (0.97) −0.57	1.28 (1.23) 2.65**	0.87 (0.88) −0.95	0.82 (0.78) −5.99***	0.70 (0.65) −5.06***	0.72 (0.38) −1.98
[0, +60]	0.33 (0.25) −25.58***	0.33 (0.26) −24.23***	0.47 (0.37) −9.43***	0.92 (0.88) −1.40	0.83 (0.77) −2.56**	1.06 (0.69) 0.27	0.72 (0.64) −4.84***	1.06 (0.99) 0.78	0.64 (0.30) −4.23***	0.80 (0.78) −5.56***	0.66 (0.64) −8.46***	0.75 (0.38) −2.44**
[0, +90]	0.30 (0.25) −36.96***	0.30 (0.26) −35.23***	0.43 (0.26) −13.25***	0.97 (0.99) −0.62	0.89 (0.82) −2.10**	0.97 (0.60) −0.21	0.66 (0.60) −6.34***	1.07 (1.03) 1.20	0.57 (0.28) −6.71***	0.75 (0.78) −9.44***	0.60 (0.56) −13.84***	0.66 (0.38) −4.39***

Changes in EU-ETS liquidity in response to events are evaluated using the quoted, relative and effective bid-ask spread ratios of the EEX EUA futures contracts for December 2008 and December 2009 delivery. The ratios are constructed to compare liquidity of the selected futures contracts in the period before the events (−90) and various event windows around the event dates. The quoted bid-ask spread is the difference between the daily best ask and bid prices, the relative bid-ask spread is the daily ask price minus the best bid price divided by the daily best mid-quote, effective bid-ask spread is twice the absolute value of the prevailing transaction price minus the daily best mid-quote. The spread ratios are calculated as the ratio of the average spreads of selected contracts over the relevant event window to the average of the spreads for the pre-event period (0, −90). The mean ratios are reported along with the corresponding median ratios in parenthesis. The t-statistic is given below the values in each box. Event 1 is the transition of trading from Phase I to Phase II. Event 2 and Event 4 are dates for the release of emissions verification results for compliance years 2007 and 2008, respectively. Event 3 is the date for the adoption of European Commission regulation (EC) No 994/2008 of 8 October 2008. Event 1 and Event 2 are investigated using the December 2008 contract, and Event 3 and Event 4, the December 2009 contract. A regular t-statistic is used to test the null hypothesis that the mean of the reported ratio for the contracts is equal to one

*, ** and *** correspond to statistical significance at 10, 5 and 1% levels, respectively

significantly reduced over the 11 trading day period around the day of the introduction of Phase II EU-ETS trading. The significant spread reductions over the longer event time intervals such as [0, +60] and [0, +90] indicate that the reduction in trading costs is not reversed for up to 90 trading days after the start of Phase II trading. This implies that changes made to trading rules and the tightening of emission caps for Phase II are both associated with liquidity improvements. The significant decline in the bid-ask spread remains intact regardless of which liquidity measure employed as can be seen in Table 5.6.

Figure 5.1, showing the time series plot of the quoted, relative and effective bid-ask spread measures for the December 2008 EEX EUA futures contract from 1 June 2007 to 26 November 2008, supports this conclusion. The figure evidences a structural shift in spread estimates occurring about the start of trading in Phase II on 2 January 2008. The figure also reiterates the narrowing of the spread measures over the long term. The spread irregularity observed towards the end of the time series is normal for the EUA futures contracts when nearing maturity.

The findings thus show a statistically significant increase in the liquidity of the Dec-2008 contract after the start of trading in Phase II of the EU-ETS. In addition, it is shown that the increase in liquidity is maintained over 90 trading days of Phase II trading.

This analysis is extended to the three other events with the results presented also in Table 5.6. For Events 2 and 4, the short-term (significant) narrowing of the quoted bid-ask spread ratios suggests that the release of the emissions verification results for 2007 and 2008 compliance years on the CITL was positively received by the market. The emissions verification results for both 2007 and 2008 were net short after those of previous years 2005 and 2006 were net long. The verification results for 2005 and 2006 (not examined in this book) reveal the market was net long at 0.81 and 0.39 billion tonnes, respectively. The actual net long positions are higher than these values from the CITL because some of the previously allocated 2.183 billion tonnes each allocated for both years were held back for auction and as new entrants' reserves. Furthermore, aggregate emissions in the EU fell by 1.6 and 3.06% for 2007 and 2008, respectively. Long-term spreads however progressively widen to indicate that the liquidity improvements seen around the announcements dates are

The Liquidity Effects of Trading Carbon Financial Instruments

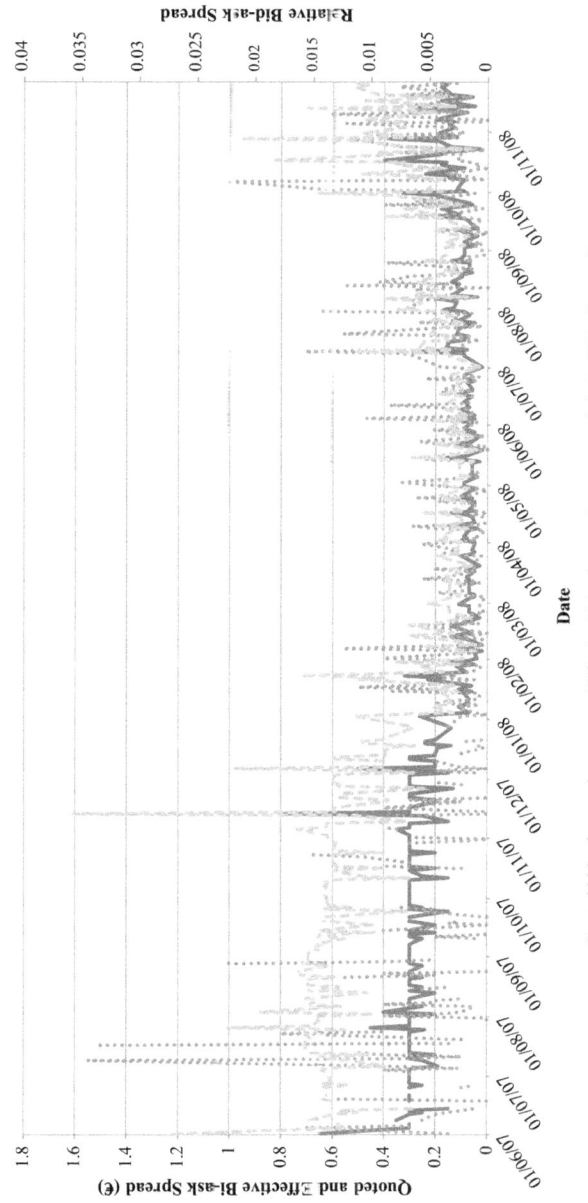

Fig. 5.1 Time series of quoted, relative and effective bid-ask spread estimates. The chart shows the daily quoted, relative and bid-ask spread estimates in Euros for the December 2008 EEX EUA Futures contract. The data spans 1 June 2007 26 November 2008

not aggressively sustained over the long term once the effects of the announcements start to wear off. This reaction sequence is not unusual in the study of events' impacts.

Short-term spread ratios for the new EC regulation (Event 3) however show that the admission by the European Commission of a need to make more secure the state of its registries about 10 months into the commencement of the Kyoto commitment phase may not have helped market confidence. The results suggest the event was associated with a loss of market liquidity in the short term. The enactment of such policy at that point in the life of an infant market has two potential interpretations. It may enhance market confidence when viewed through the lens that the policy would improve the platforms' security. Or, the policy might have indicated that the registries as they stood were not secure. Insecurity of registries indicates a dangerous development capable of jeopardising the EU climate change policy in its entirety. The latter interpretation seems to have been adopted by the market on that occasion, at least in the short term. The short-term spread estimates indicate that loss of liquidity is correlated with the timing of the announcing the new regulation. This result underscores the estimates obtained for short-term volume changes (see Table 5.4). In Table 5.4, the negative and significant estimates indicate a decrease in the trading activity and correspondingly, a decrease in the short-term liquidity is now recorded around the event period. Long-term quoted and effective spread estimates indicate that in the long term, market liquidity starts to improve significantly. This is further evidenced by results earlier reported in Table 5.5. The post/pre ratio mean (median) value of 4.72 (4.62) (and with t-statistic of 3.76) is an indication that on the long term, the policy is positively related with improving market quality hinting that market confidence finally starts to improve.

5.5 Chapter Summary

Since 2008, the world's largest mandatory carbon emissions trading scheme, the EU-ETS has moved from the first to first the second phase and now the third phase. New regulations and allocation criteria have been implemented to increase market liquidity and by extension help achieve the EU's Kyoto greenhouse gases reduction target of 8% below 1990 levels. This book is the first published study to analyse the liquidity effects on the EUA futures contracts with respect to key events and the daily evolution of transaction volumes in Phase II.

Although the EU-ETS remains the largest carbon trading scheme in the world, it is by no means a finished article; it is troubled by threats of abuses akin to conventional financial markets (see Capoor and Ambrosi 2009; Diaz-Rainey et al. 2011; Kossoy and Ambrosi 2010; Linacre et al. 2011) and consistent falling prices engendered by excess supply. The European Commission and the EU Council and Parliament are continuously drafting new policies in response to the current operational issues arising from the activities on the various platforms.

The findings made in this chapter suggest that the liquidity of EUA futures contracts is significantly enhanced when they started trading in Phase II of the EU-ETS. As the increase in liquidity spans over a 90 trading day period after the commencement of Phase II, it is suggested that there is a long-term improvement in the liquidity of the trading of carbon permits once they are traded in Phase II of the EU-ETS. This is underscored by the fact that the tested contract, the Dec-2008 EEX EUA futures contract was traded in both Phase I and Phase II of the EU-ETS. The noted liquidity improvements may be due to the regulatory changes and tighter emission caps introduced for trading in Phase II. The fact that Phase II is the Kyoto commitment period when the EU is legally bound to meet reduction obligations, may have also contributed to the observed improvements in market quality. The caps are thus tighter, resulting in rising scarcity of permits and may have been associated with improving trades volumes even on relatively less illiquid EU-ETS platforms, such as the EEX.

The rising trades' volume in Phase II does not necessarily imply improvements in liquidity; instead, structural changes such as allowing the banking of permits contribute to market confidence. Participants are more willing to trade, safe in the knowledge that the permits can be submitted for compliance purposes in the next phase (Phase III). This potentially promotes a decrease in direct trading costs and asymmetric information costs of trading (see Frino et al. 2010). Positive news on emissions verification exercises may also contribute to growing market confidence in the EU-ETS as shown by results from analysis of impacts of emissions verification results announcements in 2008 and 2009.

Evidence that the introduction of (EC) No 994/2008 of 8 October 2008 is associated with a substantial decrease of market liquidity is also reported in this chapter. The loss of liquidity on the short term may be due to a loss of market confidence stemming from a fear of insecurity at the national registries since the policy acknowledges the inadequacy of the platforms with respect to security. The policy implication is that there is a need for policymakers to ensure adequate sensitisation of market participants prior to introducing new regulations. The results show that in the long term, market liquidity is restored, thus indicating that participants altered their belief about the state of market security.

Overall, the evidence from this chapter implies that, once the caps are reliably set, uncertainty removed and trading continuity assured, a mandatory emissions trading scheme such as the EU-ETS has the potential to be successful at reducing carbon emissions.

Note

1. Initially, 42 events over a 41-month period stretching from August 2007 to December 2010 were preliminarily explored for liquidity shocks potential. From these, 12 were fully assessed; however, only four events with sufficient statistical impacts for volume and liquidity effects are reported. All the remaining eight events show from virtually no significant results.

References

Baker, K. (1996). Trading Location and Liquidity: An Analysis of U.S. Dealer and Agency Markets for Common Stocks. *Financial Markets, Institutions, and Instruments, 5*, 1–51.

Barclay, M. J., Christie, W. G., Harris, J. H., Kandel, E., & Schultz, P. H. (1999). Effects of Market Reform on the Trading Costs and Depths of Nasdaq Stocks. *The Journal of Finance, 54*, 1–34.

Beneish, M. D., & Gardner, J. C. (1995). Information Costs and Liquidity Effects from Changes in the Dow Jones Industrial Average List. *The Journal of Financial and Quantitative Analysis, 30*, 135–157.

Benz, E., & Hengelbrock, J. (2009). *Price Discovery and Liquidity in the European CO_2 Futures Market: An Intraday Analysis*. Paper presented at the Carbon Markets Workshop, 5 May 2009.

Bernstein, P. L. (1987). Liquidity, Stock Markets and Market Makers. *Financial Management, 16*, 54–62.

Branch, B., & Freed, W. (1997). Bid-Ask Spreads on the Amex and the Big Board. *The Journal of Finance, 32*, 159–163.

Brown, S. J., & Warner, J. B. (1985). Using Daily Stock Returns: The Case of Event Studies. *Journal of Financial Economics, 14*, 3–31.

Butzengeiger, S., Betz, R., & Bode, S. (2001). *Making GHG Emissions Trading Work – Crucial Issues in Designing National and International Emission Trading Systems*. Hamburg Institute of International Economics Discussion Paper 154, Hamburg.

Campbell, J. Y., Lo, A. W., & Mackinlay, A. C. (1997). *The Econometrics of Financial Markets*. Princeton, NJ: Princeton University Press.

Cao, C., Field, L. C., & Hanka, G. (2004). Does Insider Trading Impair Market Liquidity? Evidence from IPO Lockup Expirations. *The Journal of Financial and Quantitative Analysis, 39*, 25–46.

Capoor, K., & Ambrosi, P. (2009). *State and Trends of the Carbon Markets, 2009*. The World Bank Report, Washington, DC.

Carr, M. (2010). *RWE Shifts Some Carbon Trade to EEX, Curbing ECX*. London: Bloomberg Magazine Article.

Chordia, T., Roll, R., & Subrahmanyam, A. (2001). Market Liquidity and Trading Activity. *The Journal of Finance, 56*, 501–530.

Chordia, T., Roll, R., & Subrahmanyam, A. (2002). Order Imbalance, Liquidity, and Market Returns. *Journal of Financial Economics, 65*, 111–130.

Chordia, T., Roll, R., & Subrahmanyam, A. (2008). Liquidity and Market Efficiency. *Journal of Financial Economics, 87*, 249–268.

Chordia, T., & Subrahmanyam, A. (2004). Order Imbalance and Individual Stock Returns: Theory and Evidence. *Journal of Financial Economics, 72*, 485–518.

Copeland, T. E., & Galai, D. (1983). Information Effects on the Bid-Ask Spread. *The Journal of Finance, 38*, 1457–1469.

Danielsson, J., & Payne, R. (2010). *Liquidity Determination in an Order Driven Market*. London School of Economics Working Paper, London.

Daskalakis, G., Ibikunle, G., & Diaz-Rainey, I. (2011). The CO_2 Trading Market in Europe: A Financial Perspective. In A. Dorsman, W. Westerman, M. B. Karan, & Ö. Arslan (Eds.), *Financial Aspects in Energy: A European Perspective* (pp. 51–67). Berlin; Heidelberg: Springer.

Denis, D. K., McConnell, J. J., Ovtchinnikov, A. V., & Yu, Y. (2003). S&P 500 Index Additions and Earnings Expectations. *The Journal of Finance, 58*, 1821–1840.

Dennis, P., & Strickland, D. (2003). The Effect of Stock Splits on Liquidity and Excess Returns: Evidence from Shareholder Ownership Composition. *Journal of Financial Research, 26*, 355–370.

Diaz-Rainey, I., Siems, M., & Ashton, J. (2011). The Financial Regulation of Energy and Environmental Markets. *Journal of Financial Regulation and Compliance, 19*, 355–369.

Dickey, D. A., & Fuller, W. A. (1979). Distribution of the Estimators for Autoregressive Time Series with a Unit Root. *Journal of the American Statistical Association, 74*, 427–431.

Domowitz, I. (2002). Liquidity, Transaction Costs, and Reintermediation in Electronic Markets. *Journal of Financial Services Research, 22*, 141–157.

Foster, F. D., & Viswanathan, S. (1993). Variations in Trading Volume, Return Volatility, and Trading Costs: Evidence on Recent Price Formation Models. *The Journal of Finance, 48*, 187–211.

Frino, A., Kruk, J., & Lepone, A. (2010). Liquidity and Transaction Costs in the European Carbon Futures Market. *Journal of Derivatives and Hedge Funds, 16*, 100–115.

Fujimoto, A. (2004). *Macroeconomic Sources of Systematic Liquidity*. University of Alberta Working Paper, Alberta.

Glosten, L. R. (1987). Components of the Bid-Ask Spread and the Statistical Properties of Transaction Prices. *The Journal of Finance, 42*, 1293–1307.

Glosten, L. R., & Harris, L. E. (1988). Estimating the Components of the Bid/Ask Spread. *Journal of Financial Economics, 21*, 123–142.

Glosten, L. R., & Milgrom, P. R. (1985). Bid, Ask and Transaction Prices in a Specialist Market with Heterogeneously Informed Traders. *Journal of Financial Economics, 14*, 71–100.

Goyenko, R. Y., Holden, C. W., & Trzcinka, C. A. (2009). Do Liquidity Measures Measure Liquidity? *Journal of Financial Economics, 92*, 153–181.

Gregoriou, A., & Ioannidis, C. (2006). Information Costs and Liquidity Effects from Changes in the FTSE 100 List. *European Journal of Finance, 12*, 347–360.

Grossman, S. J., & Miller, M. H. (1988). Liquidity and Market Structure. *The Journal of Finance, 43*, 617–633.

Hallin, M., Mathias, C., Pirotte, H., & Veredas, D. (2011). Market Liquidity as Dynamic Factors. *Journal of Econometrics, 163*, 42–50.

Hasbrouck, J. (1991a). Measuring the Information Content of Stock Trades. *The Journal of Finance, 46*, 179–207.

Hasbrouck, J. (1991b). The Summary Informativeness of Stock Trades: An Econometric Analysis. *The Review of Financial Studies, 4*, 571–595.

Hasbrouck, J., & Schwartz, R. A. (1988). Liquidity and Execution Costs in Equity Markets. *Journal of Portfolio Management, 14*, 10–16.

Hedge, S. P., & McDermott, J. B. (2003). The Liquidity Effects of Revisions to the S&P 500 Index: An Empirical Analysis. *Journal of Financial Markets, 6*, 413–459.

Heflin, F., & Shaw, K. W. (2000). Blockholder Ownership and Market Liquidity. *The Journal of Financial and Quantitative Analysis, 35*, 621–633.

Hill, J., Jennings, T., & Vanezi, E. (2008). *The Emissions Trading Market: Risks and Challenges*. Financial Services Authority Discussion Paper, London.

Ho, T., & Stoll, H. R. (1981). Optimal Dealer Pricing under Transactions and Return Uncertainty. *Journal of Financial Economics, 9*, 47–73.

Ho, T. S. Y., & Stoll, H. R. (1983). The Dynamics of Dealer Markets under Competition. *The Journal of Finance, 38*, 1053–1074.

Johnson, T. C. (2008). Volume, Liquidity, and Liquidity Risk. *Journal of Financial Economics, 87*, 388–417.

Jones, C. M. (2002). *A Century of Stock Market Liquidity and Trading Costs*. Columbia University Working Paper, New York.

Kossoy, A., & Ambrosi, P. (2010). *State and Trends of the Carbon Markets, 2010*. The World Bank Report, Washington, DC.

Kyle, A. S. (1985). Continuous Auctions and Insider Trading. *Econometrica, 53*, 1315–1335.

Lee, C. M. C., Mucklow, B., & Ready, M. J. (1993). Spreads, Depths, and the Impact of Earnings Information: An Intraday Analysis. *The Review of Financial Studies, 6*, 345–374.

Lesmond, D. A., O'Connor, P. F., & Senbet, L. W. (2008). *Capital Structure and Equity Liquidity*. University of Auckland Working Paper, Auckland

Linacre, N., Kossoy, A., & Ambrosi, P. (2011). *State and Trends of the Carbon Market 2011*. The World Bank Report, Washington, DC.

MacKinnon, J. G. (1996). Numerical Distribution Functions for Unit Root and Cointegration Tests. *Journal of Applied Econometrics, 11*, 601–618.

Mizrach, B., & Otsubo, Y. (2014). The Market Microstructure of the European Climate Exchange. *Journal of Banking & Finance, 39*, 107–116.

Montagnoli, A., & de Vries, F. P. (2010). Carbon Trading Thickness and Market Efficiency. *Energy Economics, 32*, 1331–1336.

Newey, W. K., & West, K. D. (1987). A Simple, Positive Semi-definite, Heteroskedasticity and Autocorrelation Consistent Covariance Matrix. *Econometrica, 55*, 703–708.

Pham, P. K., Kalev, P. S., & Steen, A. B. (2003). Underpricing, Stock Allocation, Ownership Structure and Post-listing Liquidity of Newly Listed Firms. *Journal of Banking & Finance, 27*, 919–947.

Roll, R. (1984). A Simple Implicit Measure of the Effective Bid-Ask Spread in an Efficient Market. *The Journal of Finance, 39*, 1127–1139.

Sarr, A., & Lybek, T. (2002). *Measuring Liquidity in Financial Markets*. International Monetary Fund Working Paper WP/02/232, Washington, DC.

Schrand, C., & Verrecchia, R. (2005). *Disclosure Choice and Cost of Capital: Evidence from Underpricing in Initial Public Offerings*. University of Pennsylvania Working Paper, Philadelphia.

Schwarz, G. (1978). Estimating the Dimension of a Model. *The Annals of Statistics, 6*, 461–464.

Stoll, H. R. (1989). Inferring the Components of the Bid-Ask Spread: Theory and Empirical Tests. *The Journal of Finance, 44*, 115–134.

White, H. (1980). A Heteroskedasticity-Consistent Covariance Matrix Estimator and a Direct Test for Heteroskedasticity. *Econometrica, 48*, 817–838.

Lee, C. M., & Ready, M. J. (1991). Inferring Trade Direction from Intraday Data. *The Journal of Finance, 46*, 733–746.

6

Liquidity and Market Efficiency in Carbon Markets

6.1 Introduction

The extent to which the price discovery process reflects all available information in the market can be described as an indication of the market's efficiency (see Fama 1970). In regular financial markets, liquidity plays an important role in enhancing price discovery and by extension, pricing efficiency. In this chapter, we first test whether this holds for the EU-ETS and thus determine that intraday pricing efficiency is inextricably linked to daily liquidity. Specifically, we confirm that the predictability of intraday returns from intraday lagged order flows significantly decreases on days when the market enjoys greater liquidity. Fama's (1970) view of market efficiency implies the absence of return predictability, while market microstructure literature emphasises the reflection of private information in prices as a measure of market quality (see Chordia et al. 2008). Kyle (1985) notes that even the most efficient of markets reflect different levels of private information. Naturally, when markets attain higher levels of liquidity due to an exogenous event, they may more easily absorb private

---

The study described in this chapter is based on Ibikunle et al. (2016). Please see bibliography for the full reference details.

information, since increased liquidity may encourage more informed trading due to a fall in transaction costs (see Admati and Pfleiderer 1988). The link between liquidity and pricing efficiency is even more important in the case of the EU-ETS, where EU policies are being implemented in order to increase transactions in carbon permits. Such policies can be regarded as exogenous events akin to policy regarding tick size changes on NYSE, for example. In this chapter, where we identify the commencement of compliance years as an exogenous event corresponding to the tightening of trading spreads (see Fig. 6.2), a confirmation of the liquidity-return predictability link would imply that increased trading activity induced by EU policies could improve EU-ETS pricing efficiency. This is the first empirical study to directly examine the intraday evolution of this relationship in an environmental market like the EU-ETS. Second, we test whether market efficiency improves over the course of 40 months in Phase II of the EU-ETS, by using compliance years as exogenous regimes which correspond to reductions in trading spreads. The study thus presents the longest period study of intraday analysis of intraday pricing dynamics in the EU-ETS. Third, we test whether intraday prices move closer to random walk benchmarks from compliance year to compliance year in Phase II. Deviations from a random walk benchmark would implicitly suggest higher levels of noise in the trading process and vice versa. These three issues hold significance for several stakeholders. Firstly, policy makers who aim to improve trading activity as well as efficiency of the trading process may benefit from a clearer understanding of the links between return predictability and liquidity. Secondly, market makers on EU-ETS platforms, whose job it is to provide liquidity, platform operators and regulators may learn how the evolution of liquidity could affect the price discovery process. Thirdly, investors in carbon financial instruments may also find this study beneficial, as intraday prices rather than end of day prices mostly influence trader sentiment, and trades clustered around the opening/closing of the market (at which point it is most volatile—see Rotfuß 2009) may be the basis for settling derivative contracts.

As market participants require time to incorporate new information into their trading strategies, a market deemed efficient over a daily horizon does not necessarily translate into a market that is efficient at every point during the day (e.g. see Chordia et al. 2008; Epps 1979; Fama 1970; Hillmer and Yu 1979; Patell and Wolfson 1984). Confirmation of this notion is available in the contributions of Cushing and Madhavan (2000) and Chordia et al. (2005), showing that short-run returns can be predicted from order flows. However, Chordia et al. (2008) find that this predictability diminishes with improving market liquidity and across different tick size regimes on the NYSE. Similarly, Chung and Hrazdil (2010a) confirm the diminishing predictability proposition in a large sample analysis of NASDAQ stocks. These studies thus provide evidence of a strong relationship between liquidity and the enhancement of market efficiency through the impact of liquidity on the pricing process.

Another stream of literature examines the connection between liquidity and returns through the demand for *premia* when trading in illiquid instruments. Pástor and Stambaugh (2003) find a positive cross-sectional relationship between stock returns and liquidity risks. Their results are underscored by similar findings from Datar et al. (1998) and Acharya and Pedersen (2005). Similarly, Amihud (2002) documents evidence supporting the hypothesis that expected market liquidity provides an indication of stock excess return in a time series. This implies that the excess return, to some extent, typifies an illiquidity premium. Chang et al. (2010) also report consistent findings for the TSE.

Chordia et al. (2008) make an argument for the relatedness of pricing efficiency and liquidity. Consider market makers in a hypothetical market struggling to sustain liquidity supply. This may be as a result of financial difficulties or overexposure to untenable positions. Market makers may be relatively sensitive to significant buy orders, for example, or an imbalance between buy and sell orders may imply that trading is taking place on the basis of private information. When such a scenario exists, pricing strain caused by arriving order flows potentially forces a brief deviation of prices from their underlying worth (hence inefficiency; see Fama 1970). Order flow can thus give an indication of instrument returns, at least over short intervals (see also Chordia and Subrahmanyam 2004; Stoll 1978). Experienced and vigilant market participants (perhaps trading with algorithms for a significant proportion of the time) are likely to notice at least some of these deviations from random walk benchmarks, and thus become informed. Participants who remain unaware of the deviations can therefore be regarded as uninformed within the scope of those price deviations. The informed traders may tender market orders with the aim of profiting from the arbitrage opportunity. This is an informed trading activity. The choice of market orders is informed by the need to quickly profit before the arbitrage opportunity disappears, which would most likely be fleeting. The submitted orders from the arbitrageurs, assuming they are made in sufficient volumes and on time, are the ones that would lead to reduced pressure on the market makers' inventories. This then leads to the correction of the asset prices. According to Chordia et al. (2005), the correction in asset prices decreases return predictability. Since arbitrage traders (also known as informed traders—see Grossman and Stiglitz 1980) are more likely to tender orders when the spreads are narrow, one would expect reduced return predictability when the market is more liquid than otherwise (e.g. see Brennan and Subrahmanyam 1998 for the influence of liquidity on trading strategies; Peterson and Sirri 2002). We investigate this hypothesis in a unique market, the EU-ETS, created as a result of climate change policy.[1] In a market like the EU-ETS, which is set up to enable firms whose emissions are constrained under EU law (compliance traders) to exchange emission permits, the activities of arbitrageurs demand examination. Suppose arbitrageurs enter the market in order to exploit price deviations from instruments' underlying values. This may not enhance value for compliance traders, since such action may lead to failure of the scheme through the erosion of confidence. Even market makers, available in several EU-ETS platforms, may not consistently provide a sufficient level of liquidity throughout the trading day. However, if, as proposed by the foregoing hypothesis, arbitrageurs' activity enhances pricing efficiency, then ultimately compliance traders may benefit from trading in a market with a diversified pool of traders, including non-compliance informed traders. Nonetheless, it appears that the activities of informed traders are limited in the EU-ETS. Bredin et al.'s (2011) findings imply that liquidity traders dominate informed traders on the ECX, the largest trading venue by volume in the EU-ETS. Thus, informed trading activity might actually be limited enough to avoid impairing market integrity, but sufficient to enhance short-run pricing efficiency as argued by Chordia et al. (2008). A few other studies also investigate informed trading in the EU-ETS (see as examples, the study reported in Chap. 3 and Kalaitzoglou and Maher Ibrahim 2013). The aforementioned studies' findings suggest that informed trading activity in the EU-ETS is linked with improved market efficiency, as is the case in established markets.

The study reported in this chapter mainly employs the short-horizon order imbalance and return predictability regressions methodology of Chordia et al. (2008) to examine the relationship between liquidity and market efficiency. Our main findings are as follows: (1) intraday return predictability is significantly reduced when the traded instruments are relatively more liquid, hence short-horizon pricing is closely linked with liquidity—daily liquidity robustness tests and causality analyses also support this conclusion; (2) short-run instrument trading return predictability is reduced as the market evolves/matures over a 40-month period—this indicates that pricing efficiency is enhanced progressively over the same period; (3) the ECX instruments tested show improvement in terms of conformity with random walk benchmarks with each successive compliance period, thus implying a progressive reduction in noise trading. Overall, the findings suggest that the ECX has achieved a comparable level of informational efficiency to long-established financial markets. By econometrically establishing an intraday link between return predictability and liquidity in the EU-ETS, this chapter differs from the previous literature on trading activity in the EU-ETS. For example, Frino et al. (2010) and Benz and Hengelbrock (2009) investigate market liquidity, while Daskalakis (2013) and Montagnoli and de Vries (2010) investigate market efficiency in the EU-ETS. However, none of these papers examine possible links between carbon pricing and liquidity. Showing this link is important from both policy and economic perspectives. Although the findings reported in Chap. 3 suggest that carbon instruments with higher levels of liquidity (and lower information asymmetry) are priced more efficiently, those results do not constitute a direct linking of carbon pricing to liquidity. Furthermore, their paper mainly focuses on the comparison of carbon market trading activity during regular trading hours and after hours using a ten-month period data. Therefore, this study, based on a 40-month trading data, could be viewed as an extension of their work.[2]

The remainder of the chapter is arranged as follows: Sect. 6.2 discusses sample selection and describes the data. Section 6.3 reports our econometric methodology and the empirical findings, and finally Sect. 6.4 concludes.

6.2 Data

The sample period runs for most of Phase II of the EU-ETS. We compute all of the order imbalance, liquidity and futures returns measures on an instrument-specific basis. The use of instrument-specific variables for the examination of pricing dynamics is grounded in the microstructure literature (e.g. see Brennan et al. 1993; Lo and MacKinlay 1990). Serial dependence between days is approximately zero for dynamic instruments, which are too heavily traded for anomalies to exist for very long (see Chordia et al. 2005). An examination of connections between liquidity and pricing on a platform such as the ECX (with actively traded instruments) should therefore be centred on intraday trading. Serial dependence in intraday returns has already been reported by Conrad et al. (2012). The phenomenon is a result of persistence in order flow, which suggests that orders move significantly in a particular direction. If the change in direction is for buys, then the returns become positively persistent, if it is for sells, the returns will turn persistently negative. This persistence in returns will subsist, usually over short intervals; until the order flow is balanced by countervailing orders (see Chordia et al. 2005 for a comprehensive analysis of this process). The carbon futures order imbalance analysis of Mizrach and Otsubo (2014) using daily measures will not, therefore, suffice. As Mizrach and Otsubo (2014) focus on daily order imbalance, the results have no relevance for our research questions. We use 15-minute intervals as the focus of this study rather than the five-minute intervals used by other studies (see as examples Chordia et al. 2008; Chung and Hrazdil 2010a, b) based on two considerations:

1. Concerns regarding non-trading for less actively traded instruments on ECX: since on EU-ETS derivative trading platforms the nearest maturity contracts usually account for about 80% of trades in emission permits, non-trading is taken into account by extending the interval to 15 minutes.
2. Theoretically, a predictive connection between order imbalances and returns in a dynamic market should not endure for more than a matter of minutes (for less than 60 minutes according to Chordia et al. 2005), since market inefficiencies create arbitrage opportunities.

Trading activity on account of this in turn helps regain a measure of pricing efficiency. Hence, while non-trading remains an issue, the time interval chosen must be short enough to capture the inefficiencies in trading activity. This informs the decision not to extend the interval beyond 15 minutes.

Based on the foregoing and the problem of determining serial dependence in the presence of non-trading, we restrict our analysis to instruments that are traded relatively frequently for a specific compliance year.[3] We also follow Chordia et al. (2001) in excluding instruments with differing trading characteristics to the major CFIs traded on the exchange. Thus, we exclude EUA Daily futures and EUA spreads. Ultimately, only five EUA futures contracts with varying trading years are retained in the sample. These are the December expiry contracts for 2008, 2009, 2010, 2011 and 2012.[4] Crucially, these December maturity contracts account for more than 76% of daily trading volume on the ECX for the period under investigation. The dataset obtained directly from the ICE/ECX London platform comprises of all intraday tick-by-tick ECX EUA futures contracts on-screen trades on the ECX platform from January 2008 through to April 2011. The dataset contains date, timestamp, market identifier, product description, traded month, order identifier, trade sign (bid/offer), traded price, quantity traded, parent identifier, and trade type. The dataset is split into four compliance periods of Phase II, the 2008, 2009, 2010 and 2011 compliance years. Tick size for the entire period is €0.01, having been reduced from €0.05 on 27 March 2007 (during Phase I). The division based on regulatory compliance periods is exogenous and provides for a practical basis to examine the effect of liquidity on return predictability.

6.2.1 Order Imbalance and Return Measures

We begin our enquiry into the links between liquidity and pricing by computing the variables employed for most of this chapter. In the dataset, all trades are labelled as either buyer-initiated or seller-initiated, hence we do not need to algorithmically allocate trade classifications. We use two order imbalance methods based on nominal trades and the euro weight of trades, respectively. Nominal order imbalance ($OIBQ_t$) for each CFI and for each 15-minute interval is calculated using Eq. (6.1), while the euro order imbalance ($OIB€_t$) is simply the weighing of (6.1) with euro trading value of the trades, as shown in Eq. (6.2).[5] The nominal order imbalance is non-weighted and thus fails to account for the economic significance of the trades, unlike the euro order imbalance measure. For most of the analysis we therefore employ the latter measure of order imbalance.

$$OIBQ_t = \frac{(\text{Buy}_t - \text{Sell}_t)}{(\text{Buy}_t + \text{Sell}_t)} \quad (6.1)$$

$$OIB€_t = \frac{(€\text{Buy}_t - €\text{Sell}_t)}{(€\text{Buy}_t + €\text{Sell}_t)} \quad (6.2)$$

For each CFI, the measures are calculated for every 15-minute interval in a trading day. In computing the 15-minute returns variable, we use the last transaction prices for every 15-minute period.[6] Although returns are viewed as being less biased when computed from quotes than from transaction prices, we face two challenges in computing from quotes. First, heterogeneities exist for trading rates of occurrence for CFIs on the ECX as a result of the high level of differing trading frequencies. Potentially, the results will be biased and contradictory since the method could result in returns being computed over different spans for different CFIs. One should also be mindful that computing returns with transaction prices could be affected by bid-ask bounce. However, the summary statistics of trading prices for each of the four compliance periods suggests that this is not a significant source of concern in the ECX dataset. Also, since our analyses are based on contract-specific measures, we mitigate the problem of non-synchronicity. In addition, when a contract fails to trade at $t-1$, we do not use it when constructing the market aggregate measure at time t. We therefore employ the transaction prices for computing the return variable for each 15-minute period such that the return for 10:15 am is computed using the last trade at 10:00 am and the last trade at 10:15 am.

For all analyses utilising lagged estimates, the earliest 15-minute interval for each trading session/day is removed since it is connected to the lagged interval of the foregoing trading session.

6.2.2 Liquidity Measures

We adopt two measures of short-term liquidity constructed using transaction prices at regular 15-minute intervals. Relative spread, the main measure of liquidity used, is given as the best/highest traded bid price minus the best/lowest traded ask price. This is then divided by the average of the best-traded bid and best-traded ask price for every 15-minute period. Traded spread, the second measure of liquidity employed, is defined as the best-traded bid price minus the best-traded ask price over the same interval.[7] The traded spread is computed for use in robustness analysis and most of the results for it are not presented in this chapter.

Contract-specific daily bid-ask spread measures are first computed and then aggregated cross-sectionally across the CFI samples to obtain a market-wide liquidity value. Four exogenous liquidity periods are identified based on compliance periods available in Phase II of the EU-ETS. The sample thus spans 40 months of trading based on available data. Period (1) runs from 2 January 2008 until 31 December 2008; period (2) from 2 January 2009 until 31 December 2009; period (3) from 4 January 2010 until 31 December 2010; and period (4) from 3 January 2011 until 29 April 2011. Period (4) is restricted by data availability for this study. The four periods correspond to Phase II-Year I, Phase II-Year II, Phase II-Year III and Phase II-Year IV, respectively. Since they jointly represent two-thirds of Phase II, they have a good degree of representativeness.

6.3 Results and Discussion

6.3.1 Summary Statistics

Table 6.1 shows summary statistics for traded spread, relative spread, market returns and the two order imbalance measures. The samples are representative of values for all contracts examined in the assigned periods,

Table 6.1 Descriptive statistics for 15-minute liquidity and order imbalance proxies

		Traded spread (€)	Relative spread	$OIBQ_t$	$OIB€_t$
Entire sample	Mean	69.00	3.45	−18.90	−28.70
$t = 847$	Median	54.00	2.88	0.24	−21.40
$t = 58,836$	Std. Dev.	51.00	2.10	210.00	200.00
Phase II, Year I	Mean	120.00	4.31	−5.79	−82.80
$t = 256$	Median	103.00	3.75	6.12	−89.50
$t = 14,125$	Std. Dev.	58.00	2.17	330.00	250.00
Phase II, Year II	Mean	65.00	4.26	−30.20	−12.50
$t = 254$	Median	60.00	3.70	−14.10	14.80
$t = 18,550$	Std. Dev.	29.00	2.34	180.00	210.00
Phase II, Year III	Mean	34.00	2.28	−26.90	−14.20
$t = 255$	Median	31.00	2.10	−20.90	−15.70
$t = 21,239$	Std. Dev.	11.00	0.80	140.00	170.00
Phase II, Year IV	Mean	31.00	1.92	20.4	2.77
$t = 82$	Median	25.00	1.55	1.90	7.77
$t = 4922$	Std. Dev.	17.00	0.94	90.00	130.00

The table shows descriptive statistics for liquidity and order imbalance measures computed from trading data for the ECX; all presented values are multiplied by a factor of 10^3. Traded spread is the difference of best-traded bid and ask prices for every 15-minute period; relative spread is defined as the best-traded bid minus the best-traded ask price, then divided by the average of the best-traded bid and best ask over the same intervals. The spreads are averaged across the day for each instrument and cross-sectionally across all instruments. $OIBQ_t$ is the difference between the number buyer-initiated trades and seller-initiated trades, divided by total trades over every 15-minute interval in a trading day. $OIB€_t$ is given as the Euro value of buyer-initiated trades less Euro value of seller-initiated trades, then divided by the total Euro value of trades over the same interval. T is the total number of days during a trading period/year; t is the aggregate number 15-minute intervals for all the instruments. The data is for Phase II of the EU-ETS and spans 2 January 2008–29 April 2011. The data includes December maturity EUA futures contracts traded during the period. The delineation of the contracts analysed for each year, based on trading activity constraints, is as follows: for Year 2008, the December 2008, 2009 and 2010 maturities are used; for Year 2009, December 2009, 2010 and 2011 maturities are used; for Year 2010, December 2010, 2011 and 2012 maturities are used and the December 2011 and 2012 maturities are used for Year 2011.

with the exclusion of those missing observations (i.e. when no trading occurs). The statistics clearly show a strong improvement in liquidity over the four compliance periods. The mean traded and relative spread measures for Phase II-Year I are €0.1197 and 0.0043, respectively. These decrease substantially by 74 and 56% to €0.031374 and 0.001918 in Year IV for the traded and relative spreads, respectively. Order imbalance measures generally increase in magnitude, progressing from negative (−0.005787 and €−0.08279) in Year I to positive values in the Year IV (0.02042 and €0.002767) for $OIBQ_t$ and $OIB€_t$, respectively. The statistics are similar to those of Chordia et al. (2008) and thus, show a market with an increasing ratio of bid trades to ask trades. Markets with the highest level of pricing efficiency usually record more bid than ask trades, since arbitrage traders operate from neutral positions, from which they bid for opportunities.

6.3.2 Correlations

Table 6.2 shows correlation coefficients for (lags of) the two order imbalance measures and the futures returns. As one would expect, and consistent with Chordia et al. (2008), the order imbalance measures are clearly highly correlated, except for in Year I, however, all are statistically significant at the 1% level. The return is not as highly correlated with the order imbalance measures. These results are more in line with correlations between daily horizon measures, as reported in Chordia et al. (2005), where very low correlations are reported between daily order imbalance measures and returns. Although it may be surprising that we record statistically significant negative correlations for Year II, this is not unusual, especially when frequencies of less than five minutes are used in computing the return and order imbalance measures (see Chordia et al. 2005). For other periods and for the $OIBQ_t$ measure, the correlation coefficients are positive and also statistically significant. For $OIB€_t$, only Year III returns a coefficient that is not statistically significant at all levels tested. The initial expectation here is that of a high correlation of returns with the order imbalance measures. Although this is not the case, the correlation analysis results are by no means conclusive evidence of the link between returns and order imbalance on the ECX. It is important to understand that while there are already several investment vehicles on the carbon trading platform, the main aim of the market is not for wealth creation, but

Table 6.2 Correlations for 15-minute trading intervals on the ECX

		Return	$OIBQ_{t-1}$
Whole sample (2008–2011)	$OIBQ_{t-1}$	0.01	
$t = 58{,}836$	$OIB€_{t-1}$	0.00	0.53***
Phase II, Year I	$OIBQ_{t-1}$	0.02*	
$t = 14{,}125$	$OIB€_{t-1}$	0.02*	0.39***
Phase II, Year II	$OIBQ_{t-1}$	−0.01*	
$t = 18{,}550$	$OIB€_{t-1}$	−0.01*	0.80***
Phase II, Year III	$OIBQ_{t-1}$	0.02**	
$t = 21{,}239$	$OIB€_{t-1}$	0.01	0.64***
Phase II, Year IV	$OIBQ_{t-1}$	0.02	
$t = 4922$	$OIB€_{t-1}$	0.03**	0.55***

The table shows correlations values for three variables. $OIBQ_t$ is the difference between number buyer-initiated trades and seller-initiated trades, divided by total trades over every 15-minute interval in a trading day. $OIB€_t$ is given as the Euro value of buyer-initiated trades less Euro value of seller-initiated trades, then divided by the total Euro value of trades over the same interval. The return is computed for every 15-minute interval using the last trade at every interval. t is the aggregate number of 15-minute intervals for all the instruments. The data is for Phase II of the EU-ETS and spans 2 January 2008–29 April 2011. The data includes December maturity EUA futures contracts traded during the period. ***, ** and * represent statistical significance at 1, 5 and 10% levels, respectively. The delineation of the contracts analysed for each year, based on trading activity constraints, is as follows: for Year 2008, the December 2008, 2009 and 2010 maturities are used; for Year 2009, December 2009, 2010 and 2011 maturities are used; for Year 2010, December 2010, 2011 and 2012 maturities are used and the December 2011 and 2012 maturities are used for Year 2011

rather to establish an emission-constrained economy. This complicates the picture for the market efficiency-liquidity relationship we are investigating.

6.3.3 Predictive Regressions, Market Efficiency and Liquidity

We employ Chordia et al. (2008) returns predictability Model (6.3) to estimate the level of short-horizon efficiency. If Ret_t is the contract return and Order Imbalance$_{t-1}$ corresponds to either of $OIBQ_t$ or $OIB€_t$ during the previous 15-minute interval, then;

$$Ret_t = \alpha + \beta_1 \text{ Order Imbalance}_{t-1} + \varepsilon_t \qquad (6.3)$$

Based on results of the correlation analysis, the expectations for the predictive regressions are not very clear. Table 6.3 reports results for the predictive regressions of 15-minute returns on lag order imbalance measures. Panel A shows the results for regressions run with the $OIBQ_t$ measure and Panel B for the $OIB\mathcal{E}_t$ measure. The results are not stable across the various periods for both regressions. For the regressions run with the entire sample, the results suggest that on the ECX, lagged order imbalance is not a significant predictor for short-run returns and this is consistent for both order imbalance measures. However, different findings are obtained for the period-based results. For example, in Panel A, for Phase II, Year I, Phase II, Year III and Phase II, Year IV, the $OIBQ_t$ coefficients

Table 6.3 Predictive regressions of 15-minute returns on lagged order imbalance

Dependent variable: Ret_t		
Whole sample (2008–2011)	Coefficient	t-statistic
Panel A		
Intercept	0.35	0.13
$OIBQ_{t-1}$	18.10	1.46
R-squared		3.60
Phase II, Year I		
Intercept	−5.09	−0.84
$OIBQ_{t-1}$	32.50*	1.77
R-squared		25.30
Phase II, Year II		
Intercept	−1.23	−0.19
$OIBQ_{t-1}$	0.36	1.00
R-squared		24.40
Phase II, Year III		
Intercept	2.69	1.19
$OIBQ_{t-1}$	36.80**	2.32
R-squared		22.20
Phase II, Year IV		
Intercept	7.54**	2.08
$OIBQ_{t-1}$	44.1**	2.10
R-squared		5.40

(continued)

Table 6.3 (continued)

Dependent variable: Ret_t		
Whole sample (2008–2011)	Coefficient	t-statistic
Panel B		
Intercept	−0.17	−0.07
$OIB€_{t-1}$	5.90	0.46
R-squared		0.40
Phase II, Year I		
Intercept	−8.28**	−2.34
$OIB€_{t-1}$	0.58**	2.07
R-squared		22.90
Phase II, Year II		
Intercept	−0.84	−0.14
$OIB€_{t-1}$	54.7*	1.84
R-squared		18.30
Phase II, Year III		
Intercept	1.96	0.88
$OIB€_{t-1}$	0.18	1.36
R-squared		8.70
Phase II, Year IV		
Intercept	−1.67	−0.26
$OIB€_{t-1}$	0.44*	1.80
R-squared		8.68

Panel A shows results for 15-minute predictive regressions for EUA Futures contracts on the ECX. $OIBQ_{t-1}$ is the difference between number buyer-initiated trades and seller-initiated trades, divided by total trades over the 15-minute interval, $t-1$. The Ret_t, computed for every 15-minute interval using the last trade at every interval is the return at interval, t. Panel B shows results for 15-minute predictive regressions for EUA Futures contracts on the ECX. $OIB€_{t-1}$ is given as the Euro value of buyer-initiated trades less Euro value of seller-initiated trades, then divided by the total Euro value of trades over the 15-minute interval, $t-1$. The Ret_t, computed for every 15-minute interval, using the last trade at every interval, is the return at interval, t. Specifically, the following regression is estimated using least squares with Newey and West (1987) HAC:

$$Ret_t = \alpha + \beta_1 \text{ Order Imbalance}_{t-1} + \varepsilon_t$$

All coefficient and R^2 values are multiplied by a factor of 10^5. ***, ** and * represent statistical significance at 1, 5, and 10% levels, respectively. The data used is for Phase II of the EU-ETS and spans 2 January 2008–29 April 2011. The data includes December maturity EUA futures contracts traded during the period. The delineation of the contracts analysed for each year, based on trading activity constraints, is as follows: for Year 2008, the December 2008, 2009 and 2010 maturities are used; for Year 2009, December 2009, 2010 and 2011 maturities are used; for Year 2010, December 2010, 2011 and 2012 maturities are used and the December 2011 and 2012 maturities are used for Year 2011 (significance-values) are 0.000325 (0.08), 0.000368 (0.02), and 0.000441 (0.03), respectively. The R^2 for the two periods are 0.025, 0.022, and 0.0054%, respectively. Similarly, and more significantly, in four of the periods considered using the $OIB\text{€}_t$ measure, Panel B reports statistically significant coefficients for three of the periods. In Years I, II and IV of Phase II trading, the $OIB\text{€}_t$ coefficients (significance levels) are 0.000582 (5%), 0.000547 (10%), and 0.000436 (10%), respectively. The R^2 values for the three periods are, respectively, 0.023, 0.02, and 0.009%. Thus, if the overall estimates for the 40-month period examined are not considered, we can see the influence of order imbalances in the determination of market returns. It is evident that this explanatory power decreases with each passing period, which is consistent with Chordia et al. (2008). These results indicate that return predictability decreases over the period under examination, and thus, as argued by Chordia et al. (2008), the reduction in return predictability enhances pricing efficiency.

Considering the trading frequency on the ECX and the intervals of 15-minutes examined, these values are substantial and provide a basis to explore further the hypothesis that lagged order imbalance influences short-run returns. Since there is a suggestion of period dependency to the estimates, we examine next how pricing has evolved over the four periods. We run monthly predictive regressions for the 40 months under observation, starting with January 2008 and ending with April 2011. For this purpose and subsequently, we use the $OIB\text{€}_t$ as the sole order imbalance measure, since it represents the economic significance of the trading imbalance and its results in Table 6.3 are more significant. We expect that as the market's pricing efficiency improves, the power of order imbalance in predicting short-run returns diminishes; hence R^2 values are expected to drop progressively over the entire period. Figure 6.1 shows a plot of the monthly regressions R^2 values and t-statistics for the entire period. The plot shows the R^2, decreasing from a height of 1.4831% in September 2008 to 0.0004% in February 2011. The t-statistic values have remained largely stable and positive, especially over the April 2009– April 2011 period. The values hit a peak of 3.81 in February 2008 and end with a value of 1.08 in April 2011.

In keeping with Chordia et al. (2008) and Chung and Hrazdil (2010a), we next examine the connections between liquidity and

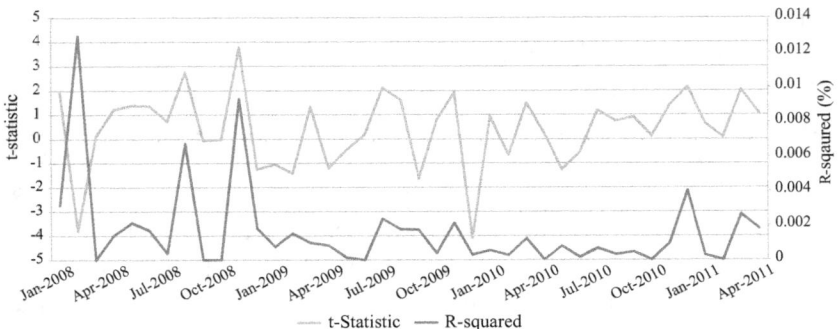

Fig. 6.1 Market efficiency measured by 15-minute return predictions with 15-minute lagged Euro Order Imbalance.
The figure shows the plot of predictive R^2s and corresponding t-statistics obtained from monthly regressions using 15-minute predictive regressions for EUA Futures contracts on the ECX. Order Imbalance $€_{t-1}$ equals the Euro value of buyer-initiated trades less Euro value of seller-initiated trades, then divided by the total Euro value of trades over the 15-minute interval, $t-1$. The Ret_t computed for every 15-minute interval using the last trade at every interval is the return at interval, t. The following regression is thus estimated:

$$Ret_t = \alpha + \beta_1 \text{ Order Imbalance}_{t-1} + \varepsilon_t$$

The data is for Phase II of the EU-ETS and spans 2 January 2008–29 April 2011. The data includes December maturity EUA futures contracts traded during the period. The delineation of the contracts analysed for each year, based on trading activity constraints, is as follows: for Year 2008, the December 2008, 2009 and 2010 maturities are used; for Year 2009, December 2009, 2010, and 2011 maturities are used; for Year 2010, December 2010, 2011 and 2012 maturities are used and the December 2011 and 2012 maturities are used for Year 2011. Note that none of the R^2 values is actually zero

market efficiency, which is the main object of this study. The most straightforward means to proxy liquidity is to examine the bid-ask spread. Traded spread is therefore measured as the highest bid minus the lowest ask price. A variant of this (see Amihud and Mendelson 1986) is the relative spread. It is defined as the traded spread divided by the average of the bid and ask prices. Although liquidity is clearly an important component of the cost of trading, Amihud (2002) suggests an alternative that is argued to more directly proxy liquidity. In less liquid markets, any given level of trading volume will give rise to a greater price response than in liquid markets. The Amihud (2002) ratio is therefore defined as the ratio of the absolute return to trading volume. However, trading volume is likely to be greater for economically larger instruments, thus creating a large firm bias. For robustness, we therefore also adopt the Florackis et al. (2011) measure, in which volume in the Amihud (2002) ratio is replaced by the turnover ratio. The principle is similar, in that a greater price movement is anticipated for illiquid markets for any given proportion of the asset traded. The advantage of this measure over the Amihud (2002) ratio is that there is no significant correlation between instrument trading size and the turnover ratio. The Amihud (2002) and Florackis et al. (2011) ratios are given as Eqs. (6.4) and (6.5) respectively:

$$\text{Amihud}_{it} = \frac{1}{D_{it}} \sum_{d=1}^{D_{it}} \frac{|R_{itd}|}{V_{itd}} \tag{6.4}$$

$$\text{Florackis}_{i} = \frac{1}{D_{it}} \sum_{d=1}^{D_{it}} \frac{|R_{itd}|}{TR_{itd}}, \tag{6.5}$$

where R_{itd}, V_{itd} and TR_{itd} are the return, euro volume and turnover ratio of EUA futures contract i on day d at month t, and D_{it} is the number of trading days in month t for EUA futures i. Larger values therefore indicate that the market is less liquid.

Figure 6.2 presents the time series plot of relative and traded spread over the 40-month period under observation.[8] As can be seen, both measures of liquidity are consistent. Conspicuously, there is a sustained narrowing of the market spread over the entire period, suggesting that liquidity improves with time on the ECX.[9] Furthermore, there seems to be a noticeable blip around the start of trading for each period. This shows a temporary loss of liquidity, perhaps present as a result of some form of calendar effect.[10]

The results for the price impact ratios are presented in Fig. 6.3. It can be seen that the evolutions of the measures are consistent with Fig. 6.2.

In reference to Blume et al. (1989) and Cox and Peterson (1994), the impact of illiquidity is more significant during periods of very low liquidity.

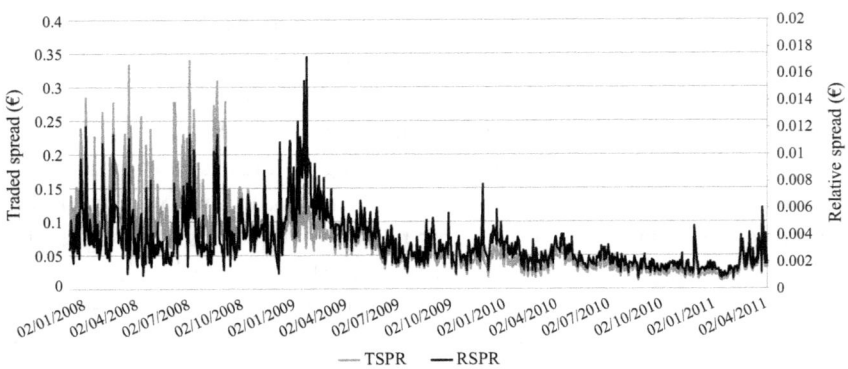

Fig. 6.2 Daily average traded spread and relative spread for ECX, 2008–2012. The figure shows the plot of daily average of liquidity measures, traded spread and relative spread. Traded spread (TSPR) is the difference of best-traded bid and ask prices for every 15-minute period; relative spread (RSPR) is defined as the best-traded bid minus the best-traded ask price, then divided by the average of the best-traded bid and best ask over the same intervals. The spreads are averaged across the day for each instrument and cross-sectionally across all instruments. The data is for Phase II of the EU-ETS and spans 2 January 2008–29 April 2011. The data includes December maturity EUA futures contracts traded during the period. The delineation of the contracts analysed for each year, based on trading activity constraints, is as follows: for Year 2008, the December 2008, 2009 and 2010 maturities are used; for Year 2009, December 2009, 2010 and 2011 maturities are used; for Year 2010, December 2010, 2011 and 2012 maturities are used and the December 2011 and 2012 maturities are used for Year 2011

In view of this, and in order to maintain consistency with Chordia et al. (2008), we split the days in our sample based on those that are considered liquid and those that are relatively illiquid. The illiquid days are defined as those whose average relative spread for that day is at least one standard deviation above the mean relative spread for a surrounding period over $(-30, +30)$.[11]

In Table 6.4, we present descriptive statistics for relative spreads on high- and low-liquidity days over the 40-month period. The general trend is that the number of illiquid days, as a proportion of the total number of trading days in the respective periods, decreases from year to year. In Phase II, Year I; Phase II, Year II; Phase II, Year III and Phase II, Year IV,[12] for example, the proportion of illiquid days as percentages of total number of trading days are 13.58; 12.89; 10.87 and 9.33%, respectively.

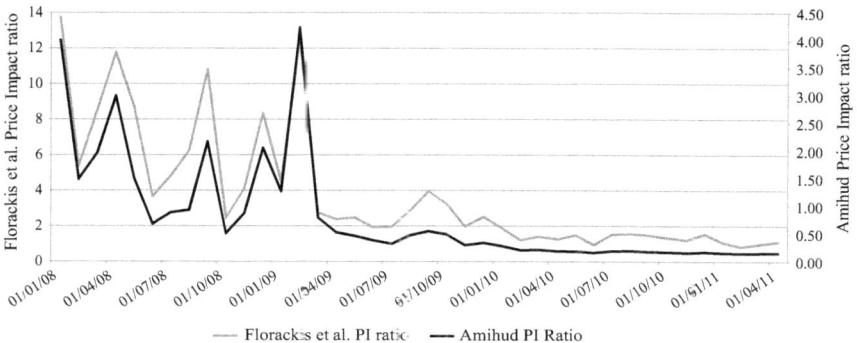

Fig. 6.3 Monthly illiquidity/price impact ratios.
The figure shows the plot of monthly computed illiquidity ratios of Amihud (2002) and Florackis et al. (2011), respectively, given below as (i) and (ii):

$$\text{Amihud}_{it} = \frac{1}{D_{it}} \sum_{d=1}^{D_{it}} \frac{|R_{itd}|}{V_{itd}} \quad \text{(i)}$$

$$\text{Florackis}_{it} = \frac{1}{D_{it}} \sum_{d=1}^{D_{it}} \frac{|R_{itd}|}{TR_{itd}}, \quad \text{(ii)}$$

where R_{itd}, V_{itd} and TR_{itd} are the return, euro volume and turnover ratio of EUA futures contract i on day d at month t, and R_{it} is the number of trading days in month t for EUA futures i. The ratios are averaged cross-sectionally across all instruments. The data is for Phase II of the EU-ETS, and spans 2 January 2008–29 April 2011. The data includes December maturity EUA futures contracts traded on the ECX during the period. The delineation of the contracts analysed for each year, based on trading activity constraints, is as follows: for Year 2008, the December 2008, 2009 and 2010 maturities are used; for Year 2009, December 2009, 2010 and 2011 maturities are used; for Year 2010, December 2010, 2011 and 2012 maturities are used and the December 2011 and 2012 maturities are used for Year 2011

The gradual improvement evident here is consistent with Figs. 6.2 and 6.3. If considered in tandem with Fig. 6.1, there is a suggestion that market efficiency improves when the market is more liquid.[13] We next test for this empirically.

In order to empirically analyse the effect of liquidity on the evolution of market efficiency, we interact the order imbalance variable ($OIB€_t$)

Table 6.4 Distribution of liquid and illiquid days on the ECX (2008–2011)

		Relative spread on liquid days	Relative spread on illiquid days
Phase II, Year I	Mean	4.31	4.37
	Median	3.74	4.45
	Std. Dev.	2.22	0.85
	Sample days, T	223	33
Phase II, Year II	Mean	4.17	4.98
	Median	3.61	4.51
	Std. Dev.	2.36	2.08
	Sample days, T	225	29
Phase II, Year III	Mean	2.29	2.22
	Median	2.09	2.10
	Std. Dev.	8.24	0.55
	Sample days, T	230	25
Phase II, Year IV	Mean	1.94	1.65
	Median	1.59	1.42
	Std. Dev.	0.97	0.52
	Sample days, T	75	7

The table shows distribution of liquid and illiquid days on the ECX. The daily relative spread, measure of liquidity, is computed using December maturity EUA futures contracts. The relative spread is defined as the best-traded bid minus the best-traded ask price scaled by the average of the best-traded bid and ask prices over 15-minute intervals, then averaged for each trading day. For every day when the daily relative spread is at least one standard deviation above the mean relative spread for the surrounding period over (−30, +30), we define the day as illiquid and when it is not, liquid. All spread estimates are multiplied by a factor of 10^3. The data is for Phase II of the EU-ETS and spans 2 January 2008–29 April 2011. The data includes December maturity EUA futures contracts traded during the period. The delineation of the contracts analysed for each year, based on trading activity constraints, is as follows: for Year 2008, the December 2008, 2009 and 2010 maturities are used; for Year 2009, December 2009, 2010 and 2011 maturities are used; for Year 2010, December 2010, 2011 and 2012 maturities are used and the December 2011 and 2012 maturities are used for Year 2011

with a low-liquidity day dummy. For every day on which the average relative spread is at least one standard deviation above the mean relative spread for a surrounding period over (−30, +30), the dummy takes on the value of 1 and 0 otherwise. If ILD_t corresponds to the low-liquidity dummy on day t, then Eq. (6.3) becomes;

$$Ret_t = \alpha + \beta_1 \text{ Order Imbalance}_{t-1} + \beta_2 \text{ Order Imbalance}_{t-1}{}^* ILD + \varepsilon_t \quad (6.6)$$

The regression Eq. (6.6) is thus run such that on low-liquidity days, the lagged $OIB€_t$ variable with the dummy interaction becomes $OIB€_{t-1}$ and 0 on other days.[14] In Table 6.5, we present the results for the regression analysis. In this analysis, we focus on the significance of the interaction term,

Table 6.5 Predictive regressions of 15-minute returns on lagged $OIB€_t$ and lagged $OIB€_t$ interacted with an illiquidity dummy

		Coefficient	t-statistic	Probability	
Phase II, Year I	Intercept	−0.09	−0.14	0.89	
t = 13,869	$OIB€_{t-1}*ILD$	33.40***	4.05	0.00	
	$OIB€_{t-1}$	1.62	0.64	0.52	
	R-squared				38.00
Phase II, Year II	Intercept	−0.07	−0.12	0.91	
t = 18,296	$OIB€_{t-1}*ILD$	33.10***	3.80	0.00	
	$OIB€_{t-1}$	3.08	0.98	0.33	
	R-squared				13.90
Phase II, Year III	Intercept	0.20	0.89	0.37	
t = 20,984	$OIB€_{t-1}*ILD$	4.38	1.12	0.26	
	$OIB€_{t-1}$	1.23	0.86	0.39	
	R-squared				1.45
Phase II, Year IV	Intercept	0.82**	2.31	0.02	
t = 4840	$OIB€_{t-1}*ILD$	20.70**	2.23	0.03	
	$OIB€_{t-1}$	1.91	0.64	0.52	
	R-squared				4.51

The table shows results for 15-minute predictive regressions for EUA Futures contracts on the ECX. $OIB€_{t-1}$ is given as the Euro value of buyer-initiated trades less Euro value of seller-initiated trades, then divided by the total Euro value of trades over the 15-minute interval, $t-1$. The Ret_t is computed for every 15-minute interval using the last trade at every interval is the return at interval, t. The dummy ILD is 1.0 for every day when the daily relative spread is at least one standard deviation above the mean relative spread for the surrounding period over (−30, +30) and 0 otherwise. Specifically, the following regression is estimated using least squares with Newey and West (1987) HAC:

$$Ret_t = \alpha + \beta_1 \text{ Order Imbalance}_{t-1} + \beta_2 \text{ Order Imbalance}_{t-1} * ILD + \varepsilon_t$$

All coefficient and R^2 values are multiplied by a factor of 10^4. ***, ** and * represent statistical significance at 15 and 10% levels respectively. The data is for Phase II of the EU-ETS and spans 2 January 2008–29 April 2011. The data includes December maturity EUA futures contracts traded during the period. The delineation of the contracts analysed for each year, based on trading activity constraints, is as follows: for Year 2008, the December 2008, 2009 and 2010 maturities are used; for Year 2009, December 2009, 2010 and 2011 maturities are used; for Year 2010, December 2010, 2011 and 2012 maturities are used and the December 2011 and 2012 maturities are used for Year 2011 coefficient β_2. The coefficients for three of the examined periods are positive and statistically significant. For Phase II, Year I, the coefficient is 0.003341 (p-value = 0.00). Also for Phase II, Year II, the coefficient is 0.003313 (p-value = 0.00). The trend holds also for the final year under consideration, Phase II-Year IV, with a coefficient of 0.0021 (p-value = 0.03). As in Panel B of Table 6.3, the Phase II, Year III period does not conform to the general trend observed for the other periods. The highest coefficient value is recorded for Year I, and decreases from then on until a slight rise in Year IV. One reason that could be advanced for the deviation of Phase II, Year III (i.e. 2010) is the inability of the United Nation's Conference of Parties to reach a highly expected and ambitious agreement on global climate policy in December 2009 at Copenhagen. The failure to achieve the expected goals set for the Copenhagen conference thus appear to have impacted the carbon markets in subsequent months.

The coefficients on lagged order imbalance (β_1) are all statistically insignificant. This implies that where lagged order imbalance is found to affect/predict returns, it does so only when the market is illiquid. Results in Table 6.5 show that this finding on illiquidity and lagged order flow explains a greater proportion of variation in returns. Consistent with NASDAQ and NYSE evidence by Chung and Hrazdil (2010b) and Chordia et al. (2008), respectively, the R^2 values decrease from Year I to Year IV. For example, in Year I, the R^2 is 0.0038, but by Year IV, this has diminished to 0.000451. This observed decline in the explanatory power of the periodic models and their significance are consistent with the gradual improvement in liquidity reported in Figs. 6.2 and 6.3, as well as in Table 6.1. Adjustments in liquidity thus affect levels of market efficiency. The liquidity improvements over 15-minute periods, which lead to enhanced pricing efficiency, may be connected with the activities of arbitrageurs as reported by Kalaitzoglou and Maher Ibrahim (2013). The reported dominant role of liquidity traders in the EU-ETS, as reported by Bredin et al. (2011), is important for the provision of counter-parties for arbitrageurs when order imbalance occurs or intensifies.

Exogenous impacts can lead to extreme order imbalances, which can in turn result in a loss of liquidity. Thus in our analysis, using the low-liquidity dummy on $OIB\epsilon_t$ to determine the impact of liquidity, rather than capturing the effect of liquidity in improving market efficiency, we may well have been picking up the impact of exogenous shocks. In order to examine this possibility we therefore construct absolute order imbalances for liquid and illiquid periods following the method of Chordia et al. (2008). We construct these measures for all four of the periods and find only minimal variations across both the liquid and illiquid days. This result implies that our results indeed capture the real role of liquidity in improving market efficiency, along with several other checks mentioned in the footnotes. As discussed by Chordia et al. (2008), it is illogical to assume that illiquidity is jointly determined with signed order imbalances. If an illiquidity-inducing event that is exogenous to the market does occur, one should expect to see as fewer buy orders as sell orders.

6.3.4 Granger Causality Analyses

As an additional check of robustness and to further explore the notion that narrowing spreads on day t results in improved pricing on day $t + 1$, we carry out Granger causality tests (see Granger 1969) relating past values of the liquidity proxy to order imbalance.[15] This is undertaken by estimating a vector autoregressive model (VAR), in this case with two dependent variables. The first, φ, is the daily relative spread liquidity proxy. The second, δ, denotes the daily mean of the 15-minute Euro order imbalance measure for each EUA futures contract traded. The lag length, l, is chosen to purge autocorrelation. This results in a model of the following form:

$$\varphi_t = \alpha_0 + \alpha_1 \varphi_{t-1} + \ldots + \alpha_l \varphi_{t-l} + \beta_1 \delta_{t-1} + \ldots + \beta_l \delta_{t-l} + \varepsilon_t$$
$$\delta_t = \alpha_0 + \alpha_1 \delta_{t-1} + \ldots + \alpha_l \delta_{t-l} + \beta_1 \varphi_{t-1} + \ldots + \beta_l \varphi_{t-l} + \mu_t \quad (6.7)$$

The equations in Model (6.7) are estimated simultaneously for each of the four periods. The statistics presented in Table 6.6 are the Wald statistics for the joint hypothesis:

$$\beta_1 = \beta_2 = \ldots = \beta_l = 0, \quad (6.8)$$

Table 6.6 Liquidity and market efficiency dynamics: Granger causality analysis

		Order imbalance does not cause relative spread	Relative spread does not cause order imbalance
Panel A. Daily contract-specific (year-dependent) liquidity and market efficiency dynamics			
Phase II, Year I	Dec-2008 futures	1.48	2.40***
	Dec-2009 futures	0.97	2.93***
	Dec-2010 futures	0.78	2.04*
Phase II, Year II	Dec-2009 futures	1.98**	2.63***
	Dec-2010 futures	5.24***	2.77***
	Dec-2011 futures	1.56	2.88***
Phase II, Year III	Dec-2010 futures	1.06	2.44**
	Dec-2011 futures	1.70**	1.71**
	Dec-2012 futures	4.69***	2.54**
Phase II, Year IV	Dec-2011 futures	0.19	0.14
	Dec-2012 futures	0.30	0.98

	Order imbalance does not cause relative spread	Relative spread does not cause order imbalance
Panel B. Daily contract-specific liquidity and market efficiency dynamics: Granger causality analysis		
Dec-2008 futures	1.48	2.40***
Dec-2009 futures	1.50	4.76***
Dec-2010 futures	1.68*	4.45***
Dec-2011 futures	0.65	3.39***
Dec-2012 futures	2.83***	2.26***

The table shows results for the Granger (1969) causality analysis. A model of the following form:

$$\varphi_t = \alpha_0 + \alpha_1 \varphi_{t-1} + \ldots \alpha_l \varphi_{t-l} + \beta_1 \delta_{t-1} + \ldots \beta_l \delta_{t-l} + \varepsilon_t$$
$$\delta_t = \alpha_0 + \alpha_1 \delta_{t-1} + \ldots \alpha_l \delta_{t-l} + \beta_1 \varphi_{t-1} + \ldots \beta_l \varphi_{t-l} + \mu_t$$

is estimated for the likely pairings of (δ, φ) series in the set. δ denotes the daily mean of the 15-minute Euro order imbalance measure, φ is the daily relative

(continued)

Table 6.6 (continued)

spread liquidity proxy and *l* is the lag length. *F*-statistics in the table are the Wald statistics for the joint hypothesis:

$$\beta_1 = \beta_2 = \ldots = \beta_l = 0$$

for each of the bivariate equations. The null tested is that δ does not Granger-cause φ in the top equation and that φ does not Granger-cause δ in the bottom equation. Panel A shows contract-specific (year/period-dependent) results, while Panel B shows contract-specific results with no breaks. ***, ** and * represent statistical significance at 1, 5 and 10% levels, respectively. The data is for Phase II of the EU-ETS and spans 2 January 2008–29 April 2011. The variables are computed from December maturity EUA futures contracts traded during the period. The delineation of the contracts analysed for each year, based on trading activity constraints, is as follows: for Year 2008, the December 2008, 2009 and 2010 maturities are used; for Year 2009, December 2009, 2010 and 2011 maturities are used; for Year 2010, December 2010, 2011 and 2012 maturities are used and the December 2011 and 2012 maturities are used for Year 2011

for each of the equations in (6.7). The null tested is that δ does not Granger-cause φ in the top equation and that φ does not Granger-cause δ in the bottom equation. We run these equations on a contract-specific basis for each period; the results are presented in Panel A of Table 6.6.

Panel A shows that the hypothesis that *liquidity improvements on a given day do not inform pricing during the next* is strongly rejected for all tested contracts for the first three periods. For the fourth period, which includes trading data for only 82 days of trading, we cannot reject this hypothesis. There is also evidence of two-way causation with respect to the Dec-2009 (Phase II, Year II), Dec-2010 (Phase II, Year III), Dec-2011 (Phase II, Year III) and the Dec-2012 (Phase II, Year III) EUA futures contracts. Broadly, therefore, the evidence of a two-way causation cannot be conclusively established since it affects contracts only in specific periods. Further, the same contracts are not affected uniformly across all periods in which they are tested. In order to explore further the possibility of bidirectional causality in the futures contracts, we also run the VAR on a contract-specific basis without recourse to sub-periods. The results are presented in Panel B of Table 6.6. Again, there is substantial evidence of Granger causality running from liquidity to order imbalance. However, a two-way causation is only observed in the case of the Dec-2012 contract. The evidence in Table 6.6 therefore further confirms our overriding hypothesis that liquidity on emissions permit trading platforms results in more efficient trading.

6.3.5 Variance Ratios: Measuring Randomness of Returns

A key assertion in this chapter is that the predictability of intraday (short-run) return from order imbalances (order flow) is an inverse indicator of pricing efficiency. Chordia et al. (2008) propose another procedure involving the measuring of randomness of returns series. This requires the comparison of variance ratios over short and long horizons. The variances of long-horizon returns are divided by the estimated variance for returns over shorter intervals. For a market in harmony with the random walk process, the variance of returns measured over longer horizons is equal to the sum of variances of shorter horizon returns as long as the summation of the shorter horizons is equal to that of the longer horizon. Based on our earlier results, we maintain the hypothesis that, during liquid spells, there will be fewer deviations from random walk benchmarks, that is, the variance ratios will be closer to one for each instrument. We concede the point made by Grossman and Miller (1988) that divergence from a random walk can be induced by inventory-related issues due to return serial correlation. However, in a largely efficient market arbitrage opportunities created by this deviation will lure participants into providing the required liquidity, hence the divergence from a random walk will be very much temporary, even if market makers cannot absorb orders (see also Admati and Pfleiderer 1988).

For each instrument within each period, we compute 15-minute returns as described in Sect. 6.3 and also for opening to closing returns for all trading days. We then compute the variance ratio by multiplying the 15-minute return variance by the number of 15-minute periods in a trading day. In a market conforming to a random walk process the variance ratio values will be close to one. Table 6.7 presents the result of the variance ratio analysis for contracts traded in each period, as well as value-weighted variance ratios for each period. The weight employed is the total trading

Table 6.7 Daily variance ratios for EUA futures contracts across trading periods

	Year I	Year II	Year III	Year IV
Dec-2008	2.18			
Dec-2009	9.47	3.46***		
Dec-2010	16.61	3.13***	1.08***	
Dec-2011		5.68	1.96***	1.43*
Dec-2012			2.44	1.37 **
Value weighted overall	3.74	3.22	1.39	1.42

The table shows contract-specific ratio of 15-minute return variance to open-to-close return variance, scaled by the number of 15-minute intervals in a trading day on the ECX. The final row in the table is the trading value-weighted aggregate variance ratio per period. Data is as defined as Table 6.1. ***, ** and * denote values which are significantly different from the previous compliance year's value at 1, 5 and 10% levels, respectively. The data used is for Phase II of the EU-ETS and spans 2 January 2008–29 April 2011. The variables are computed from December maturity EUA futures contracts traded during the period. The delineation of the contracts analysed for each year, based on trading activity constraints, is as follows: for Year 2008, the December 2008, 2009 and 2010 maturities are used; for Year 2009, December 2009, 2010 and 2011 maturities are used; for Year 2010, December 2010, 2011 and 2012 maturities are used and the December 2011 and 2012 maturities are used for Year 2011

value per instrument. As expected, consistent with Fig. 6.2 and Table 6.1, the overall variance ratio values decrease progressively from 3.74 in Year I to 1.39 in Year III. These values rise slightly to 1.42 in first four months of Year IV. It is therefore evident that as the market becomes more liquid and the EUA futures contracts are traded faster with minimal impact on their prices, the market approaches a variance ratio of unity; that is, it conforms more to the random walk process. Given that the variance ratio analysis implicitly assumes that deviations from the random walk benchmark reflects higher levels of noise in the trading process, the results suggest that noise trading generally decreased from 2008 to 2011 in the EU-ETS. A decrease in trading-related noise also implies higher market quality, and thus, we confirm earlier results that pricing/trading efficiency has improved within the sample period. Another trend in Table 6.7 is the propensity for the least traded contracts during a specific period to have the highest variance ratios. For example, in Year I, the Dec-2010 contract is the least traded and has a ratio of 16.61, more than seven times the value for the most traded contract in Year I (Dec-2008). The ratio disparity is, however, less severe in later years. For example, the Dec-2012 ratio in Year III is only about twice the value of that of the Dec-2010 contract. The results therefore support the expectation that divergence from random walk benchmarks is less severe if markets are liquid.

6.4 Chapter Summary

The EU-ETS is the largest emissions trading experiment in the world. Although there are now about 17 other compulsory schemes elsewhere in the world, the success of the EU-ETS would still provide a strong impetus for the formulation of a truly global market-driven climate change policy given that it drives more than 90% of global carbon trading. However, to the best of our knowledge, no study has been undertaken on the contemporaneous linkages between pricing efficiency and liquidity in the EU-ETS. Our work focuses on the link between efficiency and liquidity and hence contributes to the understanding of this market and provides evidence of its efficacy. Theoretically, returns predictability on an efficient market should be momentary and infrequent, and in the event it occurs, arbitrageurs should provide the necessary pool for order absorption.

In this chapter, we investigate the predictability of returns from intraday order flow across 40 months of trading on the world's largest emissions trading platform. We provide evidence that while return predictability occurs on the platform, it has significantly decreased since the start of Phase II in 2008 and continues to decline over the entire period investigated. The prices of each instrument are thus closer to the random walk benchmark as the scheme evolves.

Our results are consistent with previous studies which have investigated separately both liquidity and market efficiency. For example, Frino et al. (2010) find that aggregate long-term liquidity improves over the course of Phase I and during the early months of Phase II. In Chap. 5, we see liquidity improvements on account of enhanced regulations early on in Phase II, while Benz and Hengelbrock (2009) also report an improvement in quarterly measures of liquidity over time in Phase I. With respect to market efficiency of the EU-ETS in Phase II, our results are in line with Daskalakis (2013) and Montagnoli and de Vries (2010) as well as the findings in Chap. 3. Indeed the results of most of the existing studies investigating either the liquidity or efficiency of carbon financial instruments are similar to our findings. However, one recent study on European carbon market efficiency that appears inconsistent with the general view is that of Charles et al. (2013). Their suggestion that the European carbon market in Phase II might have been *inefficient* is related to their approach to estimating and defining market efficiency, both of which involve estimating whether spot and futures instruments in the EU-ETS conform to the cost of carry relationship.

Although the aforementioned studies examine market liquidity or efficiency during various trading phases of the EU-ETS, none links market efficiency to market liquidity. Furthermore, these studies do not provide the level of intraday exposition we have presented in this chapter, since we examine pricing and liquidity dynamics over 15-minute periods. This level of analysis is important to several stakeholders, especially carbon instruments' investors, who are most active across the trading day. This is because intraday prices usually have the most influence on trader sentiment, and thus intraday trading activity better captures market dynamics.

Based on our results and the assumption that the loss of market efficiency leads to arbitrage openings, one can conclude that arbitrage activities, especially when the market is more liquid, lead to more efficient pricing. Hence, we provide the first evidence that for emissions permit trading platforms, liquidity enhances pricing efficiency. The importance of our study is further underscored by the fact that the cost of hedging increases when efficiency decreases in futures markets (see Krehbiel and Adkins 1993). Therefore, the findings in this chapter have significant implications for more than just academic stakeholders. A number of policy makers and key stakeholders around the world remain sceptical of the effectiveness of market-led approaches in reducing GHG emissions. By testing the pricing efficiency of the ECX over a 40-month period, we provide a starting point to examine the validity of this scepticism. Policy makers (including departments responsible for the environment, energy policy, fiscal policy, trade policy and competitiveness) around the world could be reassured that improving market liquidity improves the financial instrument pricing capacity of carbon markets. This understanding could be crucial for negotiations on future global climate change legislations.

Our research also has implications for practitioners who must trade on or via EU-ETS platforms in order to comply with EU Climate Change legislations. For example, compliance traders can benefit from a stronger understanding of the carbon market dynamics. This understanding can be employed in formulating carbon investment strategies and effective carbon risk management.

While this chapter investigates the evolving relationship between market liquidity and efficiency over an extensive period in the EU-ETS, our approach, based on intraday order imbalance, return and liquidity dynamics, means that we are unable to directly/adequately test the efficiency impact of events in the EU-ETS. This is because there is no basis to expect that liquidity is jointly determined with order imbalances. Therefore, future research could consider a different analytical approach aimed at testing the efficiency impacts of key policy and economic events. Although several papers have conducted event studies to examine the impact of events on carbon prices, none has directly investigated the effect of these events on pricing efficiency. Furthermore, the ECX from which we have sourced our data is the most liquid platform in the EU-ETS and in the world. Future research could therefore investigate return predictability in relation to liquidity on less liquid platforms as well.

Notes

1. Our approach is clearly different from the spot-futures relationship approach usually adopted for measuring futures market efficiency (e.g. see Kellard et al. 1999).
2. Kalaitzoglou and Maher Ibrahim (2013) also examine informed trading in the EU-ETS by focusing on identifying the different agents at play in the carbon market.
3. If we apply 5-minute intervals as the basis for the measures, we would be forced to examine only one contract per year for the period under consideration due to non-trading effects in the other contracts; hence, we use 15-minute intervals. In any case, we conduct a robustness analysis using only the highest volume contract, which is the nearest maturity

contract, per year. We find that sampling at 5-minute intervals leaves our inferences unchanged.
4. The delineation of the contracts analysed for each year, based on trading activity constraints, is as follows: for Year 2008, the December 2008, 2009 and 2010 maturities are used; for Year 2009, December 2009, 2010 and 2011 maturities are used; for Year 2010, December 2010, 2011 and 2012 maturities are used and the December 2011 and 2012 maturities are used for Year 2011.
5. We also examine the possibility that our order imbalance measures may reflect exogenous shocks by computing absolute values for liquid and low liquidity periods as described in page 259 of Chordia et al. (2008) for robustness. As reported in Sect. 5.3, the results show that there is very low variation across the liquid and illiquid days; hence, the results are not affected by exogenous impact.
6. We also use the mid-point of the last bid and ask transaction prices for every 15-minute period with similar outcomes.
7. We also compute the measures with traded bid and ask prices at the stroke of each 15-minute period. The results are very similar with no material variation in values.
8. Further, we employ the traded spread in all other sections of this chapter for robustness examinations. In all instances, the results yielded are not materially different from those yielded by our use of the relative spread. The results presented in this chapter are thus robust to substitute liquidity proxies.
9. Given the period examined in this chapter, it is possible that the tapering off of the effects of the global financial crisis might have influenced liquidity and the enhancement of pricing efficiency during the Phase II of the EU-ETS. However, this consideration has no implication for the main focus of this study—the intraday links between return predictability and liquidity. This is because there is no basis to expect that illiquidity is jointly determined with signed order imbalances (see also Chordia et al. 2008).
10. In Chap. 5, using the market model (Brown and Warner 1985), we examine calendar effects on the European Energy Exchange (EEX) carbon platform and find no significant effect. We nevertheless apply several dummy regressions to capture the effects of specific dates in further analyses. As discussed in Sect 5.3, the dummy coefficients and respective t-statistics imply that the events have no intraday impact on our results.

11. We also employ the (⁻60, +60) window as used by Chung and Hrazdil (2010a), with no substantial differences in the number of liquid and illiquid days. For robustness, we econometrically estimate effective half-spread for days in a randomly selected month in each year using the Huang and Stoll (1997) spread decomposition model. The distribution of liquid/illiquid days observed is not qualitatively dissimilar to the one we use above.
12. In Year IV, the seven illiquid days have lower spread values than the liquid days. This underscores the limitations to our definition of illiquid days, which we control by using another window and seeing very small qualitative differences in the distribution of illiquid days. Suppose we have a period of low spreads leading to April (when compliance traders must submit emission permits) as a result of perceived increase in trading activity, and suppose most of our illiquid days are around this period. Then we are likely to obtain 'illiquid' days with lower average spread than the average of the vastly larger number of liquid days. This is plausible since our definition of illiquidity is relative to only a given number of surrounding days in the year for every illiquid day. Moreover, the differences in the values are so small that they are not statistically different from one another, even at the 10% level. The same case is made for the mean values in Year III.
13. The percentage proportion of median relative spread on illiquid days to liquid days for Years I, II, III, and IV are 119, 125, 101, and 90%, respectively. These values (except for Year IV for which we have data for only 1/3 of the year) is consistent with the expectation that the average spread for low liquidity days will be higher than that of liquid days.
14. We also carry out robustness tests by including dummies for specific day-after events such as the submission of annual emission reports, national allocation plans (NAP) and annual emission results announcements. We include dummies for days preceding specific holidays in the United Kingdom. The resultant coefficients show that the events are not significant to our investigations on intraday basis; also our earlier results are not materially altered.
15. The structure of the VAR analysis in this section is different from that of the preceding analysis, where we aim to capture return predictability, which could only be adequately investigated on an intraday basis. Conducting the VAR analysis using daily frequency data following the aggregation of 15-minute interval data into daily measures should have no implication for the results other than to reduce the probability of

establishing a statistical relationship. This is because a reduction in the degrees of freedom afforded by the use of daily aggregates of our liquidity and order imbalance measures should make it harder to establish a statistical relationship. Therefore, the results we obtain only serve to underscore the fact that past values of liquidity proxies do indeed help to explain variations in pricing.

References

Acharya, V. V., & Pedersen, L. H. (2005). Asset Pricing with Liquidity Risk. *Journal of Financial Economics, 77*, 375–410.
Admati, A., & Pfleiderer, P. (1988). A Theory of Intraday Patterns: Volume and Price Variability. *The Review of Financial Studies, 1*, 3–40.
Amihud, Y. (2002). Illiquidity and Stock Returns: Cross-Section and Time-Series Effects. *Journal of Financial Markets, 5*, 31–56.
Amihud, Y., & Mendelson, H. (1986). Asset Pricing and the Bid-Ask Spread. *Journal of Financial Economics, 17*, 223–249.
Benz, E., & Hengelbrock, J. (2009). *Price Discovery and Liquidity in the European CO_2 Futures Market: An Intraday Analysis*. Paper presented at the Carbon Markets Workshop, 5 May 2009.
Blume, M. E., Mackinlay, A. C., & Terker, B. (1989). Order Imbalances and Stock Price Movements on October 19 and 20, 1987. *The Journal of Finance, 44*, 827–848.
Bredin, D., Hyde, S., & Muckley, C. (2011). *A Microstructure Analysis of the Carbon Finance Market*. University College Dublin Working Paper, Dublin.
Brennan, M. J., & Subrahmanyam, A. (1998). The Determinants of Average Trade Size. *The Journal of Business, 71*, 1–25.
Brennan, M. J., Jegadeesh, N., & Swaminathan, B. (1993). Investment Analysis and the Adjustment of Stock Prices to Common Information. *The Review of Financial Studies, 6*, 799–824.
Brown, S. J., & Warner, J. B. (1985). Using Daily Stock Returns: The Case of Event Studies. *Journal of Financial Economics, 14*, 3–31.
Chang, Y. Y., Faff, R., & Hwang, C.-Y. (2010). Liquidity and Stock Returns in Japan: New Evidence. *Pacific-Basin Finance Journal, 18*, 90–115.
Charles, A., Darné, O., & Fouilloux, J. (2013). Market Efficiency in the European Carbon Markets. *Energy Policy, 60*, 785–792.

Chordia, T., & Subrahmanyam, A. (2004). Order Imbalance and Individual Stock Returns: Theory and Evidence. *Journal of Financial Economics, 72*, 485–518.

Chordia, T., Roll, R., & Subrahmanyam, A. (2001). Market Liquidity and Trading Activity. *The Journal of Finance, 56*, 501–530.

Chordia, T., Roll, R., & Subrahmanyam, A. (2005). Evidence on the Speed of Convergence to Market Efficiency. *Journal of Financial Economics, 76*, 271–292.

Chordia, T., Roll, R., & Subrahmanyam, A. (2008). Liquidity and Market Efficiency. *Journal of Financial Economics, 87*, 249–268.

Chung, D. Y., & Hrazdil, K. (2010a). Liquidity and Market Efficiency: A Large Sample Study. *Journal of Banking & Finance, 34*, 2346–2357.

Chung, D. Y., & Hrazdil, K. (2010b). Liquidity and Market Efficiency: Analysis of NASDAQ Firms. *Global Finance Journal, 21*, 262–274.

Conrad, C., Rittler, D., & Rotfuß, W. (2012). Modeling and Explaining the Dynamics of European Union Allowance Prices at High-Frequency. *Energy Economics, 34*, 316–326.

Cox, D. R., & Peterson, D. R. (1994). Stock Returns Following Large One-Day Declines: Evidence on Short-Term Reversals and Longer-Term Performance. *The Journal of Finance, 49*, 255–267.

Cushing, D., & Madhavan, A. (2000). Stock Returns and Trading at the Close. *Journal of Financial Markets, 3*, 45–67.

Daskalakis, G. (2013). On the Efficiency of the European Carbon Market: New Evidence from Phase II. *Energy Policy, 54*, 369–375.

Datar, V. T., Naik, N. Y., & Radcliffe, R. (1998). Liquidity and Stock Returns: An Alternative Test. *Journal of Financial Markets, 1*, 203–219.

Epps, T. W. (1979). Comovements in Stock Prices in the Very Short Run. *Journal of the American Statistical Association, 74*, 291–298.

Fama, E. F. (1970). Efficient Capital Markets: A Review of Theory and Empirical Work. *The Journal of Finance, 25*, 383–417.

Florackis, C., Gregoriou, A., & Kostakis, A. (2011). Trading Frequency and Asset Pricing on the London Stock Exchange: Evidence from a New Price Impact Ratio. *Journal of Banking and. Finance, 35*, 3335–3350.

Frino, A., Kruk, J., & Lepone, A. (2010). Liquidity and Transaction Costs in the European Carbon Futures Market. *Journal of Derivatives and Hedge Funds, 16*, 100–115.

Granger, C. W. J. (1969). Investigating Causal Relations by Econometric Models and Cross-Spectral Methods. *Econometrica, 37*, 424–438.

Grossman, S. J., & Miller, M. H. (1988). Liquidity and Market Structure. *The Journal of Finance, 43*, 617–633.

Grossman, S. J., & Stiglitz, J. E. (1980). On the Impossibility of Informationally Efficient Markets. *The American Economic Review, 70*, 393–408.

Hillmer, S. C., & Yu, P. L. (1979). The Market Speed of Adjustment to New Information. *Journal of Financial Economics, 7*, 321–345.

Huang, R. D., & Stoll, H. R. (1997). The Components of the Bid-Ask Spread: A General Approach. *The Review of Financial Studies, 10*, 995–1034.

Ibikunle, G., Gregoriou, A., Hoepner, A. G. F., & Rhodes, M. (2016). Liquidity and Market Efficiency in the World's Largest Carbon Market. *The British Accounting Review, 48*, 431–447.

Kalaitzoglou, I., & Maher Ibrahim, B. (2013). Does Order Flow in the European Carbon Futures Market Reveal Information? *Journal of Financial Markets, 16*, 604–635.

Kellard, N., Newbold, P., Rayner, T., & Ennew, C. (1999). The Relative Efficiency of Commodity Futures Markets. *Journal of Futures Markets, 19*, 413–432.

Krehbiel, T., & Adkins, L. C. (1993). Cointegration Tests of the Unbiased Expectations Hypothesis in Metals Markets. *Journal of Futures Markets, 13*, 753–763.

Kyle, A. S. (1985). Continuous Auctions and Insider Trading. *Econometrica, 53*, 1315–1335.

Lo, A., & MacKinlay, A. (1990). When Are Contrarian Profits Due to Stock Market Overreaction? *The Review of Financial Studies, 3*, 175–205.

Mizrach, B., & Otsubo, Y. (2014). The Market Microstructure of the European Climate Exchange. *Journal of Banking & Finance, 39*, 107–116.

Montagnoli, A., & de Vries, F. P. (2010). Carbon Trading Thickness and Market Efficiency. *Energy Economics, 32*, 1331–1336.

Newey, W. K., & West, K. D. (1987). A Simple, Positive Semi-Definite, Heteroskedasticity and Autocorrelation Consistent Covariance Matrix. *Econometrica, 55*, 703–708.

Pástor, L., & Stambaugh, R. F. (2003). Liquidity Risk and Expected Stock Returns. *The Journal of Political Economy, 111*, 642–685.

Patell, J. M., & Wolfson, M. A. (1984). The Intraday Speed of Adjustment of Stock Prices to Earnings and Dividend Announcements. *Journal of Financial Economics, 13*, 223–252.

Peterson, M., & Sirri, E. (2002). Order Submission Strategy and the Curious Case of Marketable Limit Orders. *The Journal of Financial and Quantitative Analysis, 37*, 221–241.

Rotfuß, W. (2009). *Intraday Price Formation and Volatility in the European Union Emissions Trading Scheme*. Centre for European Economic Research (ZEW) Working Paper, Manheim.

Stoll, H. R. (1978). The Supply of Dealer Services in Securities Markets. *The Journal of Finance, 33*, 1133–1151.

7

The Future

7.1 Policy Discussion

The studies contained in this book raise three key issues with regards to policy EU-ETS regulation and policy making.

7.1.1 Design of Regulations

In Chap. 5, we demonstrate how the European Commission's adoption of a new regulation (Commission Regulation (EC) No 994/2008) on 8 October 2008 became associated with an incremental loss of market liquidity for approximately 3 months. This example perhaps demonstrates how new regulations may unnerve a relatively nascent market like the EU-ETS. Thus, new regulations should be carefully modelled in order to avoid reversing the successes already made. The fact that this change to the existing rules was introduced just 10 months into the Kyoto commitment phase suggests that the Commission's new regime of rules was not adequate. Inadequacies in regulations may lead to loopholes in the market that can be exploited by participants. This scenario creates market uncertainty, which may contribute to a loss of market quality.

Expectedly, participants would act to protect themselves by withholding trades, and liquidity providers will impose wider spreads in order to protect themselves against informed trading.

The evidence provided by this book suggests that just like any other market, the EU-ETS does not react positively to uncertainties. The EU-ETS may be maturing quickly, but it is still a market in its infancy and must be treated as such. Policy makers must recognise this and act accordingly by showing restraint when new regulations are contemplated mid-way through a trading phase. The anticipated impacts of regulations need to be modelled before being introduced into the market. This will help in reviewing regulations in such a way that they act to forestall unwanted impacts. Ideally, the design of a trading phase should incorporate all necessary regulations. Section 7.1.2 discusses the most far-reaching mid-phase regulatory intervention in the EU-ETS until date.

7.1.2 Dealing with Excess Emission Allowances in Phase III: Backloading and the Market Stability Reserve

The European Commission aimed for the EU-ETS to deliver about 2.8 billion tonnes of emission reductions when compared with the business as usual projections over 2008–2020. However, unforeseen effects of Europe's sovereign debt crisis-induced recession means that emissions in the EU are expected to decline by as much as 3.5 billion tonnes over the same period, without the expected impact of the EU-ETS (Morris et al. 2013). In 2008, the value of EU-ETS trades stood at US$101.49 billion (€74.56 billion), which represents an 87% growth rate on the previous year and with more than three billion EUA spot, futures and option contracts traded. However, the recession in Europe subsequently led to a significant reduction in the demand for major commodities such as automobiles and real estate development projects. This in turn adversely affected the demand for raw materials used in the production of such consumer-driven products, that is, resulting in a huge fall in the demand for building and manufacturing materials such as cement and steel. The fall in general consumer demand for non-essential items thus inevitably led to a fall in industrial and consumer energy consumption. And since the need to purchase emission allowances is driven by energy consumption, the demand for EUAs fell sharply as well. Within an eight-month period, the spot price of EUAs declined 75% from a level of €28.73 in July 2008 to €7.96 on 12 February 2009.

The fall in the price of the EUAs underscores that EU-ETS platforms are efficient pricing mechanisms for carbon allowances. This is because most of the actual emissions reductions in the EU over the past decade or thereabouts have been largely due to recessionary effects and over supply of allowances. Thus, the fall in EUA price reflects a glut of allowances in the market. In July 2013, the price of EUA fell further, trading for just €4.37, a fall of more than 85% from €30 it traded for in 2006. In addition, 2013 was the first time since EU-ETS trading began that the value of emission allowances traded declined. Specifically, the value of traded carbon in the EU-ETS declined by about 35% to €62 billion at the end of 2012. Nevertheless, the resilience of the scheme was underscored by a 28% growth in the volume of traded carbon financial instruments. Overall, it is quite clear that the depression in the prices of emission allowances is due to the significant decline in European industrial production activities, which has spurred the dumping of surplus allowances on the market. Thus, if the glut can be addressed effectively relative scarcity will be restored and prices can start to appreciate again. Much higher prices are required to create the impetus for firm-level investments in low-carbon technology and processes. With the price in 2017 continuing to hover around €7.00, despite a series of policy initiatives rolled out by the European Commission to arrest the falling prices in the European carbon market, firms will continue to postpone investments in low-carbon initiatives. The EU has thus far developed two main policy initiatives to address the excess allowances in the system; these are (1) backloading of auctions in Phase III and (2) the planned deployment of the Market Stability Reserve (MSR) in 2019. Both measures correspond to short and long-term solutions to the problem, respectively.

Backloading is a short-term measure, which basically involves the postponement of the planned auctioning of 900 million emissions allowances during Phase III until 2019–2020. Thus, backloading does not alter the number of allowances auctioned during Phase II, as previously set out in the European Commission's guidance on the EU-ETS, but it simply redistributes to the volume auction across the phase. Specifically, the auctioned volumes were reduced by 400, 300 and 200 million allowances in 2014, 2015 and 2016, respectively. The Commission's impact assessment suggests that backloading can rebalance the supply and demand of emission allowances such that price volatility subsides in the short term, without undue influence on market competitiveness. Based on the foregoing, the Commission deployed its backloading programme via an amendment to the policy regulating the auctioning of emission allowances in the EU-ETS. The Commission has always acknowledged that the backloading measure is a short-term fix and that a more permanent measure would need to be introduced. However, even as a short-term fix, backloading has been a failure, with prices remaining perpetually low and volatile. Prices continue to fluctuate between just above €4 and €7, spending all but one week in 2016 under €7, and closing prices in 2017 stayed below that mark until 8 September, when it breached the ceiling to close at €7.07.

The European Commission's 'long-term solution' is unlikely to fare much better. This is the so-called MSR, which should enter into force in 2019. The excess 900 million allowances withdrawn from auctioning between 2014 and 2016 will be transferred into the reserves and become part of the MSR infrastructure. Under the MSR, the number of allowances auctioned is adjusted based on the size of the surplus of emission allowances. Specifically, once surplus allowances attain 833 million in any given year, allowances equal to 12% of the accumulated surplus will be removed from auctioning and held in reserve. These allowances are not returned to the auction block until the surplus threshold falls below 400 million allowances. At this point, the allowances are returned at the rate of 100 million in each batch. Thus, the MSR preserves the phase-dependent cap of the EU-ETS. Based on this design, The European Commission expects the MSR to control the current surplus of emissions allowances and to improve the EU-ETS's resiliency in the presence of exogenous shocks by adjusting the supply of auction allowances. However, Kollenberg and Taschini (2016) show that a cap-preserving MSR is of significance only to risk-averse firms. The MSR will not influence the expected equilibrium allowance price or average price volatility for risk-neutral firms. Therefore, until structural changes sufficient to maintain the 'right' carbon prices are made, it is unlikely that the EU-ETS can meaningfully fulfil its key purpose of driving low-carbon innovation. It remains to be seen whether continental and international politics would permit such a fundamental change.

Ibikunle and Okereke (2014) argue in favour of a more structured and responsive approach to regulating the EU-ETS. Long-term targets, such as phase-long emissions caps, should be deployed alongside a responsive and non-politicised organ for making short-term policies aimed at preserving the integrity of the market. Specifically, a fiscal regulator, with powers similar to that of a conventional market regulator or a central bank, should be set up with the singular purpose of maintaining price levels that can drive low-carbon investment growth. Instruments to be deployed by this 'regulator' would be similar to those used by the likes of the European Central Bank or the Bank of England. For example, the regulator may decide to backload a given volume of auction-able emission permits in order to shore up prices or choose to auction extra permits in order to counter artificially rising prices. Thus, in this case, the regulator's strategy is to target a certain price band, as a central bank would normally adopt an inflation target.

7.1.3 Financial Regulation of the EU-ETS

In Chap. 2, we drew attention to the issue of the regulation of carbon markets and how the ambiguity in responsible agencies may contribute to regulatory confusion and to the introduction of regulations with counter-productive effects. The regulation of EUAs (spot trading) is within the remit of the European Commission and the respective member countries, where EUAs are created as records on national registries. In the EU-ETS and indeed most commodity markets, trading in derivatives outstrip spot trading. It is therefore interesting that the regulation of the carbon permits derivatives with EUAs, EUAAs and CERs as underlyings is within the remit of different agencies depending on the country where the platform is located. For example, in the United Kingdom, the Financial Conduct Authority oversees the carbon financial instruments traded on the world's largest platform, the ECX. The current layout of regulations creates an ambiguity that can potentially affect the operation of the market. There needs to be a re-ordering of the financial regulation of the market. It appears that it would be in the interest of market quality if an independent EU-wide authority could be given the sole authority to regulate the EU-ETS with no recourse to local financial authorities. A market in its infancy requires clear signals as to the intentions of the regulators, and when those signals emanate from a myriad of sources at different or undefined levels of authority, this may result in uncertainty and ultimately losses in market confidence and quality.

The release of the first set of emission verification results in Phase I of the EU-ETS demonstrates the importance of a unified regulatory framework (see Frino et al. 2010; Daskalakis et al. 2011 for discussions on this). The leaking of verification data from national authorities' sources prior to the announcements being made by the EC results in information asymmetry in the market. Information asymmetry harms market confidence and forces market participants into adopting trading strategies that are capable of undermining the goal of the EU-ETS, which is ultimately to induce an emissions-constrained economy in the EU.

In view of these issues, policy makers could devise a unified framework for overseeing the operations and regulation of the EU-ETS. This structure should be in the form of a governing entity that can be likened to the European Central Bank. The entity should be granted the legal authority to ensure the sustenance of market quality and its members should be appointed strictly on merit with no recourse to political expediency. The creation of a single authority overseeing compliance with market regulations would be a good starting point. As the market gains in complexity and sophistication, there would be a need for more specialisation of regulatory framework at the EU level. It is likely that this would lead to the creation of more agencies with more finely defined functions. The market at this nascent stage will benefit from a unified structure. Irrespective of some of the successes recorded in recent years, global action on climate change remains an uncertain prospect, and it would benefit from more EU-ETS stability with respect to regulations.

References

Daskalakis, G., Ibikunle, G., & Diaz-Rainey, I. (2011). The CO_2 Trading Market in Europe: A Financial Perspective. In A. Dorsman, W. Westerman, M. B. Karan, & Ö. Arslan (Eds.), *Financial Aspects in Energy: A European Perspective* (pp. 51–67). Berlin; Heidelberg: Springer.

Frino, A., Kruk, J., & Lepone, A. (2010). Liquidity and Transaction Costs in the European Carbon Futures Market. *Journal of Derivatives and Hedge Funds, 16*, 100–115.

Ibikunle, G., & Okereke, C. (2014). Governing Carbon through the EU-ETS: Opportunities, Pitfalls and Future Prospects. In J. Tansey (Ed.), *Carbon Governance, Climate Change and Business Transformation*. London: Taylor and Francis (Routledge).

Kollenberg, S., & Taschini, L. (2016). *Dynamic Supply Adjustment and Banking under Uncertainty: The Market Stability Reserve*. Working paper, Available at SSRN.

Morris, D., Crow, L., Elsworth, R., MacDonald, P., & Watson, L. (2013). *Drifting Toward Disaster? The ETS Adrift in Europe's Climate Efforts* (Vol. 5). London: Sandbag.

Bibliography

Acharya, V. V., & Pedersen, L. H. (2005). Asset Pricing with Liquidity Risk. *Journal of Financial Economics, 77*, 375–410.

Admati, A., & Pfleiderer, P. (1988). A Theory of Intraday Patterns: Volume and Price Variability. *The Review of Financial Studies, 1*, 3–40.

Aitken, M., & Frino, A. (1996a). The Accuracy of the Tick Test: Evidence from the Australian Stock Exchange. *Journal of Banking & Finance, 20*, 1715–1729.

Aitken, M., & Frino, A. (1996b). Execution Costs Associated with Institutional Trades on the Australian Stock Exchange. *Pacific-Basin Finance Journal, 4*, 45–58.

Alberola, E., Chevallier, J., & Chèze, B. (2008). Price Drivers and Structural Breaks in European Carbon Prices 2005–2007. *Energy Policy, 36*, 787–797.

Albrecht, J., Verbeke, T., & Clerq, M. (2004). Informational Efficiency of the US SO_2 Permit Market. *Environmental Modelling & Software, 21*, 1471–1478.

Almeida, A., Goodhart, C., & Payne, R. (1998). The Effects of Macroeconomic News on High Frequency Exchange Rate Behavior. *The Journal of Financial and Quantitative Analysis, 33*, 383–408.

Alzahrani, A. A., Gregoriou, A., & Hudson, R. (2013). Price Impact of Block Trades in the Saudi Stock Market. *Journal of International Financial Markets, Institutions and Money, 23*, 322–341.

Amihud, Y. (2002). Illiquidity and Stock Returns: Cross-Section and Time-Series Effects. *Journal of Financial Markets, 5*, 31–56.

Amihud, Y., & Mendelson, H. (1986). Asset Pricing and the Bid-Ask Spread. *Journal of Financial Economics, 17*, 223–249.

Amihud, Y., Mendelson, H., & Lauterbach, B. (1997). Market Microstructure and Securities Values: Evidence from the Tel Aviv Stock Exchange. *Journal of Financial Economics, 45*, 365–390.

Andersen, T. G., Bollerslev, T., Diebold, F. X., & Vega, C. (2003). Micro Effects of Macro Announcements: Real-Time Price Discovery in Foreign Exchange. *The American Economic Review, 93*, 38–62.

Baker, K. (1996). Trading Location and Liquidity: An Analysis of U.S. Dealer and Agency Markets for Common Stocks. *Financial Markets, Institutions, and Instruments, 5*, 1–51.

Ball, R., & Finn, F. J. (1989). The Effect of Block Transactions on Share Prices: Australian Evidence. *Journal of Banking & Finance, 13*, 397–419.

Barclay, M. J., Christie, W. G., Harris, J. H., Kandel, E., & Schultz, P. H. (1999). Effects of Market Reform on the Trading Costs and Depths of Nasdaq Stocks. *The Journal of Finance, 54*, 1–34.

Barclay, M. J., & Hendershott, T. (2003). Price Discovery and Trading after Hours. *The Review of Financial Studies, 16*, 1041–1073.

Barclay, M. J., & Hendershott, T. (2004). Liquidity Externalities and Adverse Selection: Evidence from Trading after Hours. *The Journal of Finance, 59*, 681–710.

Barclay, M. J., Litzenberger, R. H., & Warner, J. B. (1990). Private Information, Trading Volume, and Stock-Return Variances. *The Review of Financial Studies, 3*, 233–253.

Barclay, M. J., & Warner, J. B. (1993). Stealth Trading and Volatility: Which Trades Move Prices? *Journal of Financial Economics, 34*, 281–305.

Beneish, M. D., & Gardner, J. C. (1995). Information Costs and Liquidity Effects from Changes in the Dow Jones Industrial Average List. *The Journal of Financial and Quantitative Analysis, 30*, 135–157.

Benz, E., & Hengelbrock, J. (2009). *Price Discovery and Liquidity in the European CO_2 Futures Market: An Intraday Analysis*. Paper presented at the Carbon Markets Workshop, 5 May 2009.

Bernstein, P. L. (1987). Liquidity, Stock Markets and Market Makers. *Financial Management, 16*, 54–62.

Bessembinder, H., & Seguin, P. J. (1992). Futures-Trading Activity and Stock Price Volatility. *The Journal of Finance, 47*, 2015–2034.

Biais, B., Hillion, P., & Spatt, C. (1999). Price Discovery and Learning during the Preopening Period in the Paris Bourse. *The Journal of Political Economy, 107*, 1218–1248.

Blume, M. E., Mackinlay, A. C., & Terker, B. (1989). Order Imbalances and Stock Price Movements on October 19 and 20, 1987. *The Journal of Finance, 44*, 827–848.

Boemare, C., & Quirion, P. (2002). Implementing Greenhouse Gas Trading in Europe: Lessons from Economic Literature and International Experiences. *Ecological Economics, 43*, 213–230.

Böhringer, C., & Lange, A. (2005). On the Design of Optimal Grandfathering Schemes for Emission Allowances. *European Economic Review, 49*, 2041–2055.

Branch, B., & Freed, W. (1997). Bid-Ask Spreads on the Amex and the Big Board. *The Journal of Finance, 32*, 159–163.

Bredin, D., Hyde, S., & Muckley, C. (2011). *A Microstructure Analysis of the Carbon Finance Market*. University College Dublin Working Paper, Dublin.

Bredin, D., & Muckley, C. (2011). An Emerging Equilibrium in the EU Emissions Trading Scheme. *Energy Economics, 33*, 353–362.

Brennan, M. J., Jegadeesh, N., & Swaminathan, B. (1993). Investment Analysis and the Adjustment of Stock Prices to Common Information. *The Review of Financial Studies, 6*, 799–824.

Brennan, M. J., & Subrahmanyam, A. (1996). Market Microstructure and Asset Pricing: On the Compensation for Illiquidity in Stock Returns. *Journal of Financial Economics, 41*, 441–464.

Brennan, M. J., & Subrahmanyam, A. (1998). The Determinants of Average Trade Size. *The Journal of Business, 71*, 1–25.

Brock, W. A., & Kleidon, A. W. (1992). Periodic Market Closure and Trading Volume: A Model of Intraday Bids and Asks. *Journal of Economic Dynamics and Control, 16*, 451–489.

Brown, S. J., & Warner, J. B. (1985). Using Daily Stock Returns: The Case of Event Studies. *Journal of Financial Economics, 14*, 3–31.

Burtraw, D., Goeree, J., Holt, C., Myers, E., Palmer, K., & Shobe, W. (2011). Price Discovery in Emissions Permit Auctions. In R. M. Isaac & D. A. Norton (Eds.), *Research in Experimental Economics* (pp. 11–36). Bingley, UK: Emerald Group Publishing.

Butzengeiger, S., Betz, R., & Bode, S. (2001). *Making GHG Emissions Trading Work – Crucial Issues in Designing National and International Emission Trading Systems*. Hamburg Institute of International Economics Discussion Paper 154, Hamburg.

Campbell, J. Y., Lo, A. W., & Mackinlay, A. C. (1997). *The Econometrics of Financial Markets*. Princeton, NJ: Princeton University Press.

Cao, C., Field, L. C., & Hanka, G. (2004). Does Insider Trading Impair Market Liquidity? Evidence from IPO Lockup Expirations. *The Journal of Financial and Quantitative Analysis, 39*, 25–46.

Cao, C., Ghysels, E., & Hatheway, F. (2000). Price Discovery without Trading: Evidence from the Nasdaq Preopening. *The Journal of Finance, 55*, 1339–1365.

Capoor, K., & Ambrosi, P. (2009). *State and Trends of the Carbon Markets, 2009.* The World Bank Report, Washington, DC.

Carr, M. (2010). *RWE Shifts Some Carbon Trade to EEX, Curbing ECX.* London: Bloomberg Magazine Article.

Cason, T. N., & Gangadharan, L. (2003). Transactions Costs in Tradable Permit Markets: An Experimental Study of Pollution Market Designs. *Journal of Regulatory Economics, 23*, 145–165.

Cason, T. N., & Gangadharan, L. (2011). Price Discovery and Intermediation in Linked Emissions Trading Markets: A Laboratory Study. *Ecological Economics, 70*, 1424–1433.

Chakravarty, S. (2001). Stealth-Trading: Which Traders' Trades Move Stock Prices? *Journal of Financial Economics, 61*, 289–307.

Chan, K., Chung, Y. P., & Johnson, H. (1995). The Intraday Behavior of Bid-Ask Spreads for NYSE Stocks and CBOE Options. *The Journal of Financial and Quantitative Analysis, 30*, 329–346.

Chan, K. C., Christie, W. G., & Schultz, P. H. (1995). Market Structure and the Intraday Pattern of Bid-Ask Spreads for NASDAQ Securities. *The Journal of Business, 68*, 35–60.

Chan, L. K. C., & Lakonishok, J. (1993). Institutional Trades and Intraday Stock Price Behavior. *Journal of Financial Economics, 33*, 173–199.

Chan, L. K. C., & Lakonishok, J. (1995). The Behavior of Stock Prices Around Institutional Trades. *The Journal of Finance, 50*, 1147–1174.

Chang, Y. Y., Faff, R., & Hwang, C.-Y. (2010). Liquidity and Stock Returns in Japan: New Evidence. *Pacific-Basin Finance Journal, 18*, 90–115.

Charles, A., Darné, O., & Fouilloux, J. (2013). Market Efficiency in the European Carbon Markets. *Energy Policy, 60*, 785–792.

Chiyachantana, C. N., Jain, P. K., Jiang, C., & Wood, R. A. (2004). International Evidence on Institutional Trading Behavior and Price Impact. *The Journal of Finance, 59*, 869–898.

Choi, J. Y., Salandro, D., & Shastri, K. (1988). On the Estimation of Bid-Ask Spreads: Theory and Evidence. *The Journal of Financial and Quantitative Analysis, 23*, 219–230.

Chordia, T., Roll, R., & Subrahmanyam, A. (2001). Market Liquidity and Trading Activity. *The Journal of Finance, 56*, 501–530.

Chordia, T., Roll, R., & Subrahmanyam, A. (2002). Order Imbalance, Liquidity, and Market Returns. *Journal of Financial Economics, 65*, 111–130.

Chordia, T., Roll, R., & Subrahmanyam, A. (2005). Evidence on the Speed of Convergence to Market Efficiency. *Journal of Financial Economics, 76*, 271–292.

Chordia, T., Roll, R., & Subrahmanyam, A. (2008). Liquidity and Market Efficiency. *Journal of Financial Economics, 87*, 249–268.

Chordia, T., & Subrahmanyam, A. (2004). Order Imbalance and Individual Stock Returns: Theory and Evidence. *Journal of Financial Economics, 72*, 485–518.

Chou, R. K., Wang, G. H. K., Wang, Y.-Y., & Bjursell, J. (2011). The Impacts of Large Trades by Trader Types on Intraday Futures Prices: Evidence from the Taiwan Futures Exchange. *Pacific-Basin Finance Journal, 19*, 41–70.

Christiansen, A. C., & Arvanitakis, A. (2005). Price Determinants in the EU Emissions Trading Scheme. *Climate Policy, 5*, 15–30.

Chung, D. Y., & Hrazdil, K. (2010a). Liquidity and Market Efficiency: A Large Sample Study. *Journal of Banking & Finance, 34*, 2346–2357.

Chung, D. Y., & Hrazdil, K (2010b). Liquidity and Market Efficiency: Analysis of NASDAQ Firms. *Global Finance Journal, 21*, 262–274.

Ciccotello, C. S., & Hatheway, F. M. (2000). Indicating Ahead: Best Execution and the NASDAQ Preopening. *Journal of Financial Intermediation, 9*, 184–212.

Clarke, J., & Shastri, K. (2000). *On Information Asymmetry Metrics*. SSRN eLibrary, Working Paper 251938.

Conrad, C., Rittler, D., & Rotfuß, W. (2012). Modeling and Explaining the Dynamics of European Union Allowance Prices at High-Frequency. *Energy Economics, 34*, 316–326.

Conrad, J. S., Johnson, K. M., & Wahal, S. (2001). Institutional Trading and Soft Dollars. *The Journal of Finance, 56*, 397–416.

Convery, F. J. (2009). Reflections—The Emerging Literature on Emissions Trading in Europe. *Review of Environmental Economics and Policy, 3*, 121–137.

Copeland, T. E., & Galai, D. (1983). Information Effects on the Bid-Ask Spread. *The Journal of Finance, 38*, 1457–1469.

Cox, D. R., & Peterson, D. R. (1994). Stock Returns Following Large One-Day Declines: Evidence on Short-Term Reversals and Longer-Term Performance. *The Journal of Finance, 49*, 255–267.

Cushing, D., & Madhavan, A. (2000). Stock Returns and Trading at the Close. *Journal of Financial Markets, 3*, 45–67.

Danielsson, J., & Payne, R. (2010). *Liquidity Determination in an Order Driven Market*. London School of Economics Working Paper, London.

Daskalakis, G. (2013). On the Efficiency of the European Carbon Market: New Evidence from Phase II. *Energy Policy, 54*, 369–375.

Daskalakis, G., Ibikunle, G., & Diaz-Rainey, I. (2011). The CO_2 Trading Market in Europe: A Financial Perspective. In A. Dorsman, W. Westerman,

M. B. Karan, & Ö. Arslan (Eds.), *Financial Aspects in Energy: A European Perspective* (pp. 51–67). Berlin; Heidelberg: Springer.

Daskalakis, G., & Markellos, R. N. (2008). Are the European Carbon Markets Efficient? *Review of Futures Markets, 17*, 103–128.

Daskalakis, G., Psychoyios, D., & Markellos, R. N. (2009). Modeling CO_2 Emission Allowance Prices and Derivatives: Evidence from the European Trading Scheme. *Journal of Banking & Finance, 33*, 1230–1241.

Datar, V. T., Naik, N. Y., & Radcliffe, R. (1998). Liquidity and Stock Returns: An Alternative Test. *Journal of Financial Markets, 1*, 203–219.

Davies, R. J. (2003). The Toronto Stock Exchange Preopening Session. *Journal of Financial Markets, 6*, 491–516.

Denis, D. K., McConnell, J. J., Ovtchinnikov, A. V., & Yu, Y. (2003). S&P 500 Index Additions and Earnings Expectations. *The Journal of Finance, 58*, 1821–1840.

Dennis, P., & Strickland, D. (2003). The Effect of Stock Splits on Liquidity and Excess Returns: Evidence from Shareholder Ownership Composition. *Journal of Financial Research, 26*, 355–370.

Diaz-Rainey, I., Siems, M., & Ashton, J. (2011). The Financial Regulation of Energy and Environmental Markets. *Journal of Financial Regulation and Compliance, 19*, 355–369.

Dickey, D. A., & Fuller, W. A. (1979). Distribution of the Estimators for Autoregressive Time Series with a Unit Root. *Journal of the American Statistical Association, 74*, 427–431.

Dodwell, C. (2005). *EU Emissions Trading Scheme: The Government Perspective.* Paper Presented at the Business & Investors' Climate Change Conference 2005: 31/10-01/11, 2005, London.

Domowitz, I. (2002). Liquidity, Transaction Costs, and Reintermediation in Electronic Markets. *Journal of Financial Services Research, 22*, 141–157.

Domowitz, I., Glen, J., & Madhavan, A. (2001). Liquidity, Volatility and Equity Trading Costs across Countries and over Time. *International Finance, 4*, 221–255.

Easley, D., Hvidkjaer, S., & O'Hara, M. (2002). Is Information Risk a Determinant of Asset Returns? *The Journal of Finance, 57*, 2185–2221.

Easley, D., Kiefer, N. M., & O'Hara, M. (1996). Cream-Skimming or Profit-Sharing? The Curious Role of Purchased Order Flow. *The Journal of Finance, 51*, 811–833.

Easley, D., Kiefer, N. M., & O'Hara, M. (1997). One Day in the Life of a Very Common Stock. *The Review of Financial Studies, 10*, 805–835.

Easley, D., & O'Hara, M. (1987). Price, Trade Size, and Information in Securities Markets. *Journal of Financial Economics, 19*, 69–90.

Easley, D., & O'Hara, M. (1992). Time and the Process of Security Price Adjustment. *The Journal of Finance, 47*, 577–605.

Engle, R. F., & Granger, C. W. J. (1987). Co-integration and Error Correction: Representation, Estimation, and Testing. *Econometrica, 55*, 251–276.

Epps, T. W. (1979). Comovements in Stock Prices in the Very Short Run. *Journal of the American Statistical Association, 74*, 291–298.

Fama, E. F. (1970). Efficient Capital Markets: A Review of Theory and Empirical Work. *The Journal of Finance, 25*, 383–417.

Fama, E. F., & MacBeth, J. D. (1973). Risk, Return, and Equilibrium: Empirical Tests. *The Journal of Political Economy, 81*, 607–636.

Fezzi, C., & Bunn, D. (2009). Structural Interactions of European Carbon Trading and Energy Prices. *The Journal of Energy Markets, 2*, 53–69.

Flood, M. D., Huisman, R., Koedijk, K. G., & Mahieu, R. J. (1999). Quote Disclosure and Price Discovery in Multiple-Dealer Financial Markets. *The Review of Financial Studies, 12*, 37–59.

Florackis, C., Gregoriou, A., & Kostakis, A. (2011). Trading Frequency and Asset Pricing on the London Stock Exchange: Evidence from a New Price Impact Ratio. *Journal of Banking and. Finance, 35*, 3335–3350.

Foster, F. D., & Viswanathan, S. (1990). A Theory of the Interday Variations in Volume, Variance, and Trading Costs in Securities Markets. *The Review of Financial Studies, 3*, 593–624.

Foster, F. D., & Viswanathan, S. (1993). Variations in Trading Volume, Return Volatility, and Trading Costs: Evidence on Recent Price Formation Models. *The Journal of Finance, 48*, 187–211.

French, K. R., & Roll, R. (1986). Stock Return Variances: The Arrival of Information and the Reaction of Traders. *Journal of Financial Economics, 17*, 5–26.

Frino, A., Jarnecic, E., & Lepone, A. (2007). The Determinants of the Price Impact of Block Trades: Further Evidence. *Abacus, 43*, 94–106.

Frino, A., Kruk, J., & Lepone, A. (2010). Liquidity and Transaction Costs in the European Carbon Futures Market. *Journal of Derivatives and Hedge Funds, 16*, 100–115.

Fujimoto, A. (2004). *Macroeconomic Sources of Systematic Liquidity*. University of Alberta Working Paper, Alberta.

Fung, H.-G., & Patterson, G. A. (1999). The Dynamic Relationship of Volatility, Volume, and Market Depth in Currency Futures Markets. *Journal of International Financial Markets, Institutions and Money, 9*, 33–59.

Gangadharan, L. (2000). Transaction Costs in Pollution Markets: An Empirical Study. *Land Economics, 76*, 601–614.

Garbade, K. D., & Silber, W. L. (1979). Structural Organization of Secondary Markets: Clearing Frequency, Dealer Activity and Liquidity Risk. *The Journal of Finance, 34*, 577–593.

Gemmill, G. (1996). Transparency and Liquidity: A Study of Block Trades on the London Stock Exchange under Different Publication Rules. *The Journal of Finance, 51*, 1765–1790.

George, T., Kaul, G., & Nimalendran, M. (1991). Estimation of the Bid – Ask Spread and Its Components: A New Approach. *The Review of Financial Studies, 4*, 623–656.

Glosten, L. R. (1987). Components of the Bid-Ask Spread and the Statistical Properties of Transaction Prices. *The Journal of Finance, 42*, 1293–1307.

Glosten, L. R., & Harris, L. E. (1988). Estimating the Components of the Bid/Ask Spread. *Journal of Financial Economics, 21*, 123–142.

Glosten, L. R., & Milgrom, P. R. (1985). Bid, Ask and Transaction Prices in a Specialist Market with Heterogeneously Informed Traders. *Journal of Financial Economics, 14*, 71–100.

Gonzalo, J., & Granger, C. W. J. (1995). Estimation of Common Long-Memory Components in Cointegrated Systems. *Journal of Business and Economic Statistics, 13*, 27–35.

Goodhart, C. A. E., Hall, S. G., Henry, S. G. B., & Pesaran, B. (1993). News Effects in a High-Frequency Model of the Sterling-Dollar Exchange Rate. *Journal of Applied Econometrics, 8*, 1–13.

Goyenko, R. Y., Holden, C. W., & Trzcinka, C. A. (2009). Do Liquidity Measures Measure Liquidity? *Journal of Financial Economics, 92*, 153–181.

Granger, C. W. J. (1969). Investigating Causal Relations by Econometric Models and Cross-Spectral Methods. *Econometrica, 37*, 424–438.

Greene, J. T., & Watts, S. G. (1996). Price Discovery on the NYSE and the NASDAQ: The Case of Overnight and Daytime News Releases. *Financial Management, 25*, 19–42.

Gregoriou, A. (2008). The Asymmetry of the Price Impact of Block Trades and the Bid-Ask Spread. *Journal of Economic Studies, 35*, 191–199.

Gregoriou, A., & Ioannidis, C. (2006). Information Costs and Liquidity Effects from Changes in the FTSE 100 List. *European Journal of Finance, 12*, 347–360.

Grossman, S. J., & Miller, M. H. (1988). Liquidity and Market Structure. *The Journal of Finance, 43*, 617–633.

Grossman, S. J., & Stiglitz, J. E. (1980). On the Impossibility of Informationally Efficient Markets. *The American Economic Review, 70*, 393–408.

Grubb, M., & Neuhoff, K. (2006). Allocation and Competitiveness in the EU Emissions Trading Scheme: Policy Overview. *Climate Policy, 6*, 7–30.

Grüll, G., & Taschini, L. (2011). Cap-and-trade Properties under Different Hybrid Scheme Designs. *Journal of Environmental Economics and Management, 61*, 107–118.

Gwilym, O., Buckle, M., & Thomas, S. (1997). The Intraday Behaviour of Bid-Ask Spreads, Returns and Volatility for FTSE 100 Stock Index Options. *Journal of Derivatives, 4*, 20–32.

Hallin, M., Mathias, C., Pirotte, H., & Veredas, D. (2011). Market Liquidity as Dynamic Factors. *Journal of Econometrics, 163*, 42–50.

Hanemann, M. (2009). The Role of Emission Trading in Domestic Climate Policy. *The Energy Journal, 30*, 79–114.

Hansen, L. P. (1982). Large Sample Properties of Generalized Method of Moments Estimators. *Econometrica, 50*, 1029–1054.

Harris, L. (1990). Statistical Properties of the Roll Serial Covariance Bid/Ask Spread Estimator. *The Journal of Finance, 45*, 579–590.

Hasbrouck, J. (1991a). Measuring the Information Content of Stock Trades. *The Journal of Finance, 46*, 179–207.

Hasbrouck, J. (1991b). The Summary Informativeness of Stock Trades: An Econometric Analysis. *The Review of Financial Studies, 4*, 571–595.

Hasbrouck, J. (1995). One Security, Many Markets: Determining the Contributions to Price Discovery. *The Journal of Finance, 50*, 1175–1199.

Hasbrouck, J., & Schwartz, R. A. (1988). Liquidity and Execution Costs in Equity Markets. *Journal of Portfolio Management, 14*, 10–16.

Hawksworth, J., & Swinney, P. (2009). *Carbon Taxes vs Carbon Trading*. PriceWaterhouseCoopers Report, London.

He, Y., Lin, H., Wang, J., & Wu, C. (2009). Price Discovery in the Round-the-Clock U.S. Treasury Market. *Journal of Financial Intermediation, 18*, 464–490.

Hedge, S. P., & McDermott, J. B. (2003). The Liquidity Effects of Revisions to the S&P 500 Index: An Empirical Analysis. *Journal of Financial Markets, 6*, 413–459.

Heflin, F., & Shaw, K. W. (2000). Blockholder Ownership and Market Liquidity. *The Journal of Financial and Quantitative Analysis, 35*, 621–633.

Hendershott, T., Jones, C. M., & Menkveld, A. J. (2011). Does Algorithmic Trading Improve Liquidity? *The Journal of Finance, 66*, 1–33.

Hill, J., Jennings, T., & Vanezi, E. (2008). *The Emissions Trading Market: Risks and Challenges*. Financial Services Authority Discussion Paper, London.

Hillmer, S. C., & Yu, P. L. (1979). The Market Speed of Adjustment to New Information. *Journal of Financial Economics, 7*, 321–345.

Hintermann, B. (2010). Allowance Price Drivers in the First Phase of the EU ETS. *Journal of Environmental Economics and Management, 59*, 43–56.

Ho, T., & Stoll, H. R. (1981). Optimal Dealer Pricing under Transactions and Return Uncertainty. *Journal of Financial Economics, 9*, 47–73.

Ho, T. S. Y., & Stoll, H. R. (1983). The Dynamics of Dealer Markets under Competition. *The Journal of Finance, 38*, 1053–1074.

Hobbs, B. F., Bushnell, J., & Wolak, F. A. (2010). Upstream vs. Downstream CO_2 Trading: A Comparison for the Electricity Context. *Energy Policy, 38*, 3632–3643.

Holthausen, R. W., Leftwich, R. W., & Mayers, D. (1987). The Effect of Large Block Transactions on Security Prices: A Cross-Sectional Analysis. *Journal of Financial Economics, 19*, 237–267.

Holthausen, R. W., Leftwich, R. W., & Mayers, D. (1990). Large-Block Transactions, the Speed of Response, and Temporary and Permanent Stock-Price Effects. *Journal of Financial Economics, 26*, 71–95.

Hu, S. (1997). *Trading Turnover and Expected Stock Returns: The Trading Frequency Hypothesis and Evidence from the Tokyo Stock Exchange*. National Taiwan University Working Paper, Taipei.

Huang, R. D., & Stoll, H. R. (1994). Market Microstructure and Stock Return Predictions. *The Review of Financial Studies, 7*, 179–213.

Huang, R. D., & Stoll, H. R. (1997). The Components of the Bid-Ask Spread: A General Approach. *The Review of Financial Studies, 10*, 995–1034.

Ibikunle, G., Gregoriou, A., Hoepner, A. G. F., & Rhodes, M. (2016). Liquidity and Market Efficiency in the World's Largest Carbon Market. *The British Accounting Review, 48*, 431–447.

Ibikunle, G., Gregoriou, A., & Pandit, N. (2013). Price Discovery and Trading after Hours: New Evidence from the World's Largest Carbon Exchange. *International Journal of the Economics of Business, 20*, 421–445.

Ibikunle, G., Gregoriou, A., & Pandit, N. R. (2016). Price Impact of Block Trades: The Curious Case of Downstairs Trading in the EU Emissions Futures Market. *The European Journal of Finance, 22*, 120–142.

Ibikunle, G., & Okereke, C. (2014). Governing Carbon through the EU-ETS: Opportunities, Pitfalls and Future Prospects. In J. Tansey (Ed.), *Carbon Governance, Climate Change and Business Transformation*. London: Taylor and Francis (Routledge).

Jacobsen, G. D. (2011). The Al Gore Effect: An Inconvenient Truth and Voluntary Carbon Offsets. *Journal of Environmental Economics and Management, 61*, 67–78.

Jiang, C. X., Likitapiwat, T., & McInish, T. H. (2012). Information Content of Earnings Announcements: Evidence from After-Hours Trading. *Journal of Financial and Quantitative Analysis, 47*, 1303–1330.

Johnson, T. C. (2008). Volume, Liquidity, and Liquidity Risk. *Journal of Financial Economics, 87*, 388–417.

Jones, C. M. (2002). *A Century of Stock Market Liquidity and Trading Costs.* Columbia University Working Paper, New York.

Joskow, P. L., Schmalensee, R., & Bailey, E. M. (1998). The Market for Sulfur Dioxide Emissions. *The American Economic Review, 88*, 669–685.

Joyeux, R., & Milunovich, G. (2010). Testing Market Efficiency in the EU Carbon Futures Market. *Applied Financial Economics, 20*, 803–809.

Kalaitzoglou, I., & Maher Ibrahim, B. (2013). Does Order Flow in the European Carbon Futures Market Reveal Information? *Journal of Financial Markets, 16*, 604–635.

Kara, M., Syri, S., Lehtilä, A., Helynen, S., Kekkonen, V., Ruska, M., et al. (2008). The Impacts of EU CO_2 Emissions Trading on Electricity Markets and Electricity Consumers in Finland. *Energy Economics, 30*, 193–211.

Keim, D., & Madhavan, A. (1996). The Upstairs Market for Large-Block Transactions: Analysis and Measurement of Price Effects. *The Review of Financial Studies, 9*, 1–36.

Kellard, N., Newbold, P., Rayner, T., & Ennew, C. (1999). The Relative Efficiency of Commodity Futures Markets. *Journal of Futures Markets, 19*, 413–432.

Kerr, S., & Máre, D. (1998). *Transaction Costs and Tradable Permit Markets: The United States Lead Phasedown.* Motu Economic Research Working Paper, Auckland.

Kim, M., Szakmary, A. C., & Schwarz, T. V. (1999). Trading Costs and Price Discovery across Stock Index Futures and Cash Markets. *Journal of Futures Markets, 19*, 475–498.

Kling, C., & Rubin, J. (1997). Bankable Permits for the Control of Environmental Pollution. *Journal of Public Economics, 64*, 101–115.

Koch, N. (2012). *Co-movements between Carbon, Energy and Financial Markets: A Multivariate GARCH Approach.* School of Business, Economics and Social Sciences, University of Hamburg Working Paper, Hamburg.

Kossoy, A., & Ambrosi, P. (2010). *State and Trends of the Carbon Markets, 2010.* The World Bank Report, Washington, DC.

Kraus, A., & Stoll, H. R. (1972). Price Impacts of Block Trading on the New York Stock Exchange. *The Journal of Finance, 27*, 569–588.

Krehbiel, T., & Adkins, L. C. (1993). Cointegration Tests of the Unbiased Expectations Hypothesis in Metals Markets. *Journal of Futures Markets, 13*, 753–763.

Kurov, A. (2008). Information and Noise in Financial Markets: Evidence from the E-Mini Index Futures. *Journal of Financial Research, 31*, 247–270.

Kurov, A., & Lasser, D. J. (2004). Price Dynamics in the Regular and E-Mini Futures Markets. *The Journal of Financial and Quantitative Analysis, 39*, 365–384.

Kyle, A. S. (1985). Continuous Auctions and Insider Trading. *Econometrica, 53*, 1315–1335.

Labatt, S., & White, R. R. (2007). *Carbon Finance: The Financial Implications of Climate Change*. New Jersey: John Wiley & Sons.

Lakonishok, J., & Lev, B. (1987). Stock Splits and Stock Dividends: Why, Who, and When. *The Journal of Finance, 42*, 913–932.

Lee, C. M., & Ready, M. J. (1991). Inferring Trade Direction from Intraday Data. *The Journal of Finance, 46*, 733–746.

Lee, C. M. C., Mucklow, B., & Ready, M. J. (1993). Spreads, Depths, and the Impact of Earnings Information: An Intraday Analysis. *The Review of Financial Studies, 6*, 345–374.

Lesmond, D. A., O'Connor, P. F., & Senbet, L. W. (2008). *Capital Structure and Equity Liquidity*. University of Auckland Working Paper, Auckland

Lin, J., Sanger, G. C., & Booth, G. G. (1995). Trade Size and Components of the Bid-Ask Spread. *The Review of Financial Studies, 8*, 1153–1183.

Linacre, N., Kossoy, A., & Ambrosi, P. (2011). *State and Trends of the Carbon Market 2011*. The World Bank Report, Washington, DC.

Linares, P., Santos, F. J., Ventosa, M., & Lapiedra, L. (2006). Impacts of the European Emission Trading Directive and Permit Assignment Methods on the Spanish Electricity Sector. *The Energy Journal, 27*, 79–98.

Lo, A., & MacKinlay, A. (1990). When Are Contrarian Profits Due to Stock Market Overreaction? *The Review of Financial Studies, 3*, 175–205.

MacKinnon, J. G. (1996). Numerical Distribution Functions for Unit Root and Cointegration Tests. *Journal of Applied Econometrics, 11*, 601–618.

Madhavan, A., & Cheng, M. (1997). In Search of Liquidity: Block Trades in the Upstairs and Downstairs Markets. *The Review of Financial Studies, 10*, 175–203.

Madhavan, A., & Panchapagesan, V. (2000). Price Discovery in Auction Markets: A Look Inside the Black Box. *The Review of Financial Studies, 13*, 627–658.

Madhavan, A., Richardson, M., & Roomans, M. (1997). Why Do Security Prices Change? A Transaction-Level Analysis of NYSE Stocks. *The Review of Financial Studies, 10*, 1035–1064.

Mansanet-Bataller, M., Pardo, T., & Valor, E. (2007). CO_2 Prices, Energy and Weather. *The Energy Journal, 28*, 73–92.

Mansanet-Bataller, M., & Pardo Tornero, Á. (2007). *The Effects of National Allocation Plans on Carbon Markets*. University of Valencia Working Paper, Valencia.

Miclăuş, P. G., Lupu, R., Dumitrescu, S. A., & Bobircă, A. (2008). Testing the Efficiency of the European Carbon Futures Market Using the Event-Study Methodology. *International Journal of Energy and Environment, 2*, 121–128.

Mizrach, B., & Otsubo, Y. (2014). The Market Microstructure of the European Climate Exchange. *Journal of Banking & Finance, 39*, 107–116.

Montagnoli, A., & de Vries, F. P. (2010). Carbon Trading Thickness and Market Efficiency. *Energy Economics, 32*, 1331–1336.

Montgomery, W. D. (1972). Markets in Licenses and Efficient Pollution Control Programs. *Journal of Economic Theory, 5*, 395–418.

Morris, D., Crow, L., Elsworth, R., MacDonald, P., & Watson, L. (2013). *Drifting Toward Disaster? The ETS Adrift in Europe's Climate Efforts* (Vol. 5). London: Sandbag.

Nazifi, F., & Milunovich, G. (2010). Measuring the Impact of Carbon Allowance Trading on Energy Prices. *Energy & Environment, 21*, 367–383.

Neuhoff, K., Ferrario, F., Grubb, M., Gabbel, E., & Keats, K. (2006). Emission Projections 2008–2012 Versus NAPs II. *Climate Policy, 6*, 395–410.

Newey, W. K., & West, K. D. (1987). A Simple, Positive Semi-definite, Heteroskedasticity and Autocorrelation Consistent Covariance Matrix. *Econometrica, 55*, 703–708.

O'Hara, M. (2003). Presidential Address: Liquidity and Price Discovery. *The Journal of Finance, 58*, 1335–1354.

Pascual, R., Escribano, A., & Tapia, M. (2004). Adverse Selection Costs, Trading Activity and Price Discovery in the NYSE: An Empirical Analysis. *Journal of Banking & Finance, 28*, 107–128.

Pástor, L., & Stambaugh, R. F. (2003). Liquidity Risk and Expected Stock Returns. *The Journal of Political Economy, 111*, 642–685.

Patell, J. M., & Wolfson, M. A. (1984). The Intraday Speed of Adjustment of Stock Prices to Earnings and Dividend Announcements. *Journal of Financial Economics, 13*, 223–252.

Peterson, M., & Sirri, E. (2002). Order Submission Strategy and the Curious Case of Marketable Limit Orders. *The Journal of Financial and Quantitative Analysis, 37*, 221–241.

Pham, P. K., Kalev, P. S., & Steen, A. B. (2003). Underpricing, Stock Allocation, Ownership Structure and Post-listing Liquidity of Newly Listed Firms. *Journal of Banking & Finance, 27*, 919–947.

Porter, D. C., & Weaver, D. G. (1998). Post-trade Transparency on Nasdaq's National Market System. *Journal of Financial Economics, 50*, 231–252.

Reneses, J., & Centeno, E. (2008). Impact of the Kyoto Protocol on the Iberian Electricity Market: A Scenario Analysis. *Energy Policy, 36*, 2376–2384.

Rittler, D. (2012). Price Discovery and Volatility Spillovers in the European Union Emissions Trading Scheme: A High-Frequency Analysis. *Journal of Banking & Finance, 36*, 774–785.

Roll, R. (1984). A Simple Implicit Measure of the Effective Bid-Ask Spread in an Efficient Market. *The Journal of Finance, 39*, 1127–1139.

Rotfuß, W. (2009). *Intraday Price Formation and Volatility in the European Union Emissions Trading Scheme.* Centre for European Economic Research (ZEW) Working Paper, Mannheim.

Rubin, J. D. (1996). A Model of Intertemporal Emission Trading, Banking, and Borrowing. *Journal of Environmental Economics and Management, 31*, 269–286.

Saar, G. (2001). Price Impact Asymmetry of Block Trades: An Institutional Trading Explanation. *The Review of Financial Studies, 14*, 1153–1181.

Sarr, A., & Lybek, T. (2002). *Measuring Liquidity in Financial Markets.* International Monetary Fund Working Paper WP/02/232, Washington, DC.

Schennach, S. M. (2000). The Economics of Pollution Permit Banking in the Context of Title IV of the 1990 Clean Air Act Amendments. *Journal of Environmental Economics and Management, 40*, 189–210.

Schleich, J., Ehrhart, K.-M., Hoppe, C., & Seifert, S. (2006). Banning Banking in EU Emissions Trading? *Energy Policy, 34*, 112–120.

Schmalensee, R., Joskow, P. L., Ellerman, A. D., Montero, J. P., & Bailey, E. M. (1998). An Interim Evaluation of Sulfur Dioxide Emissions Trading. *The Journal of Economic Perspectives, 12*, 53–68.

Schrand, C., & Verrecchia, R. (2005). *Disclosure Choice and Cost of Capital: Evidence from Underpricing in Initial Public Offerings.* University of Pennsylvania Working Paper, Philadelphia.

Schwarz, G. (1978). Estimating the Dimension of a Model. *The Annals of Statistics, 6*, 461–464.

Sijm, J., Neuhoff, K., & Chen, Y. (2006). CO_2 Cost Pass-through and Windfall Profits in the Power Sector. *Climate Policy, 6*, 49–72.

Sijm, J. P. M., Bakker, S. J. A., Chen, Y., Harmsen, H. W., & Lise, W. (2005). *CO_2 Price Dynamics: The Implications of EU Emissions Trading for the Price of Electricity.* Energy Research Centre of the Netherlands (ECN) Working Paper ECN-C-05-081, Amsterdam.

Springer, U. (2003). The Market for Tradable GHG Permits under the Kyoto Protocol: A Survey of Model Studies. *Energy Economics, 25*, 527–551.

Stavins, R. N. (1995). Transaction Costs and Tradeable Permits. *Journal of Environmental Economics and Management, 29*, 133–148.

Stavins, R. N. (1998). What Can We Learn from the Grand Policy Experiment? Lessons from SO_2 Allowance Trading. *The Journal of Economic Perspectives, 12*, 69–88.

Stoll, H. R. (1978). The Supply of Dealer Services in Securities Markets. *The Journal of Finance, 33*, 1133–1151.

Stoll, H. R. (1989). Inferring the Components of the Bid-Ask Spread: Theory and Empirical Tests. *The Journal of Finance, 44*, 115–134.

Stoll, H. R., & Whaley, R. E. (1990). Stock Market Structure and Volatility. *The Review of Financial Studies, 3*, 37–71.

Svendsen, G. T., & Vesterdal, M. (2003). How to Design Greenhouse Gas Trading in the EU? *Energy Policy, 31*, 1531–1539.

Uhrig-Homburg, M., & Wagner, M. (2007). Derivative Instruments in the EU Emissions Trading Scheme—An Early Market Perspective. *Energy and Environment, 19*, 1–26.

Uhrig-Homburg, M., & Wagner, M. (2009). Futures Price Dynamics of CO_2 Emission Allowances: An Empirical Analysis of the Trial Period. *Journal of Derivatives, 17*, 73–88.

van Bommel, J. (2011). Measuring Price Discovery: The Variance Ratio, the R^2 and the Weighted Price Contribution. *Finance Research Letters, 8*, 112–119.

Van Ness, B. F., Van Ness, R. A., & Warr, R. S. (2001). How Well Do Adverse Selection Components Measure Adverse Selection? *Financial Management, 30*, 77–98.

Vesterdal, M., & Svendsen, G. T. (2004). How Should Greenhouse Gas Permits Be Allocated in the EU? *Energy Policy, 32*, 961–968.

White, H. (1980). A Heteroskedasticity-Consistent Covariance Matrix Estimator and a Direct Test for Heteroskedasticity. *Econometrica, 48*, 817–838.

Zhang, Y.-J., & Wei, Y.-M. (2010). An Overview of Current Research on EU ETS: Evidence from Its Operating Mechanism and Economic Effect. *Applied Energy, 87*, 1804–1814.

Index[1]

A

Abnormal returns, 132, 133, 139, 140
Acharya, V.V., 6, 167
Admati, A., 83n1
Adverse selection component, 65
Adverse selection costs, 5, 42, 53–68, 73, 101, 151
After market closes (AMC), 40–43, 45–48, 50, 51, 53, 68, 69, 71, 74, 79, 82, 83n2
After-hours trading (AHT), 39–42, 79, 97, 113
Aitken, M., 47
Alberola, E., 9
Albrecht, J., 18
Almeida, A., 40
Alzahrani, A.A., 93, 95, 100, 102, 109, 113, 119, 126n5
Ambrosi, P., 32
Amihud, Y., 6, 167, 180, 181, 183
Andersen, T. G., 40
Arbitrage, 6, 7, 18, 150, 168, 170, 175, 190, 193
Arvanitakis, A., 9
Ask prices, 55–57, 102, 108, 111, 117, 122, 135, 138, 139
Asset prices, 2, 5, 7, 168
Asymmetric information costs, 5, 129, 159
Auctioning, 22, 25, 26, 203, 204
Auction-quote driven market, 54
Augmented Dickey-Fuller test, 144
Average abnormal returns (AAR), 140, 141
Aviation sector, 34n1
Axpo Trading AG, 133

---

[1] Note: Page number followed by 'n' refer to notes.

B

Backloading, 30, 202–205
Bank of England, 205
Barclay, M.J., 39, 40, 50, 51, 69, 76, 77, 79, 82, 84n6, 93, 118
Basic regression model, 62
Before market opens (BMO), 40
Belektron d.o.o., 133
Benz, E., 7, 8, 64, 68, 131, 169, 192
Bernstein, P. L., 149
Bessembinder, H., 102
Best-traded ask price, 182, 184
Biais, B., 40, 77
Bid prices, 56, 57, 102, 108, 111, 117, 122
Bid transactions, 62
Bid-ask spreads (BAS), 53, 54, 63, 93, 102–105, 108, 110, 111, 117, 122, 132, 135, 139, 151, 153
Block purchases, 105, 112, 113, 123, 125, 126n7
Block trades/trading, 3, 4, 41, 47, 82, 91–93, 98–100, 102–108, 113, 114, 117, 119–124, 125n3
Bluenext, 8, 29, 95, 134
Blume, M. E., 181
Bredin, D., 9, 10n1, 168, 186
Brennan, M. J., 3, 4
Breusch–Godfrey serial correlation, 141
Brock, W. A., 52
Brokers, 45, 93, 125n1
Brown, S. J., 132
Bunn, D., 9, 25
Butzengeiger, S., 129
Buyer-initiated trades, 4, 180
Buy-sell indicator variable, 63

C

Campbell, J.Y., 139, 140, 151
Cao, C., 40, 132
Cap and trade design, 18, 19, 24, 32
Capital asset pricing model (CAPM), 5
Carbon dioxide (CO_2), 2, 20, 22, 25, 27, 28, 30, 33, 97, 134
Carbon exchange, 82, 94
Carbon financial instruments (CFIs), 29, 33, 46, 79, 94, 98, 129–160, 171–173, 203
  trading, 44, 123
Carbon markets, 1, 16, 31, 33, 40, 41, 43, 113, 129, 165–197
Carbon permits, 9, 28, 41, 43, 83, 130, 139, 142, 144, 159, 166
Carbon price, 9, 26–29, 33, 194, 205
Carbon spot market, 95
Carbon trading, 7, 17, 18, 33, 44, 83, 131, 134, 139, 140, 159, 175, 192
Carousel VAT evasion strategy, 30
Causality issues, 8
Central Bank, 205, 206
Central limit order book (CLOB), 125n1
Certified Emission Reduction units (CERs), 22, 31, 95, 97, 134
Ceteris paribus, 25
Chakravarty, S., 93
Chan, L.K.C, 93
Chang, Y. Y., 6, 167
Charles, A., 193
Cheng, M., 98, 113
Chicago Climate Exchange (CCX), 94

Index

Chicago Climate Futures Exchange (CCFX), 94
Chiyachantana, C.N., 109
Chordia, T., 6, 7, 167–169, 171, 175, 176, 179, 182, 186, 187, 190
Christiansen, A. C., 9
Chung, D.Y., 6, 167, 179, 186
Ciccotello, C. S., 40, 77
Clean Development Mechanism (CDM), 16, 17, 23
Climate change policy, 168, 192
Coefficient of variation, 138
Commodities, 18, 26–31, 47, 202
Community Independent Transaction Log (CITL), 22, 97, 156
Compliance buyers, 124
CompositeLiq, 138
Confidence bands, 77, 78
Conrad, C., 170
Consumer confidence, 29
Consumer energy consumption, 203
Conventional financial assets returns, 28
Convery, F. J., 7
Copenhagen conference, 186
Correlation, 50, 57–59, 64, 71, 77, 79
Correlation coefficients, 175
Counter-productive effects, 205
Cox, D. R., 181
Critical phase-dependent issues, 23–32
Cumulative abnormal return (CAR), 140
Cushing, D., 6, 167

D

Daily variance ratios, 191
Danielsson, J., 54, 130
Daskalakis, G., 8, 9, 10n1, 27, 33, 169, 192
Datar, V.T., 6, 167
Davies, R. J., 40
Day of week effects, 112–113
De Vries, F. P., 9, 131, 169, 192
Dec-2008 contracts, 131, 134, 139, 142–145, 153
Dec-2009 contracts, 139, 142, 153
Dennis, P., 132
Dependent variables, 177, 178, 187
Depth, 109, 138, 146, 148, 152
Dickey–Fuller test, 144
Discrete variables, 139

E

EC Regulation (EC), 131, 133, 141, 144, 158
EEX Carbon Index (Carbix®), 134
Effective spread, 42, 53, 62, 65, 158
Electricity generators, 34n1
Electronic trading, 133
Emission permits, 1, 2, 26
Emission Reduction Units (ERUs), 22–23
Emission-constrained economy, 176
Emissions trading, 15–34, 42, 96, 148, 158, 160, 192
  cap and trade design, 18
Emissions-constrained economy, 25
End of day (EOD) variable, 46, 98
Energy efficiency, 16
Engle, R.F., 7

Index

Environmental financial economics, 1
Environmental policy, 1, 18
Environmental Protection Agency (EPA), 18
Equilibrium-inducing transactions, 61
EU Aviation Allowances (EUAA), 134
Euro order imbalance, 172, 180, 187, 188
European Central Bank, 205, 206
European Climate Exchange (ECX), 7, 8, 39, 42–51, 53, 59, 79, 82, 91, 93–98, 101, 103, 105, 110–112, 117, 118, 123–125, 130, 131, 133, 148, 169–172, 175–177, 179, 181, 184, 193, 194, 205
    end of day (EOD) data, 46
    EUA, 43, 45, 94, 102, 171
    information absorption, 68
    trading environment on, 43
European Commodity Clearing AG (ECC), 134
European Energy Exchange (EEX), 48, 130–134, 138
    EEA, 153
    EUA, 133, 139, 143, 145, 153–159
    trading environment, 133–134
European Union Allowances (EUAs), 8, 9, 20, 22, 23, 26–28, 30, 31, 96, 97, 101, 117, 118, 122–124, 203, 205
European Union Emissions Trading Scheme (EU-ETS), 1–3, 7–10, 17–19, 33, 41–43, 47, 52, 78, 79, 83, 93, 94, 97, 98, 102, 103, 110, 118, 123–125, 129–132, 139, 144, 148, 152, 153, 158–160, 165, 166, 168–170, 173, 182, 183, 192–194, 202, 206
    carbon price in, 26–28
    financial regulation of, 205–206
    Global Financial Crisis on, 28–30
    initial allocation in phase I, 24–26
    microstructure of, 32
    phases of, 21
    regulation, design of, 201–202
    regulatory risk issues in, 30–32
    structure of, 19–23
    trading, 91, 153, 156, 203
Euros, 47, 72, 91, 101, 103, 123, 126n4, 133, 157, 172, 174, 176, 178, 180, 185
Event period, 142, 158
Event studies, 9, 139, 141, 194
Excess emission allowances, 202–205
Exchange for physical/swaps (EFP/EFS), 44, 45, 47, 51, 53, 74, 79, 82, 83, 83n2, 95, 103
Exchange trading mechanism, 57
Exchange-traded permits, 43

Fama, E.F., 79, 165
Fezzi, C., 9, 25
Financial markets liquidity, 146–147
Fiscal regulator, 205
Florackis, C., 181, 183
Foster, F.D., 3, 83n1, 130
Fragment trades, 104
Fragmentation, 119, 129, 146
Frino, A., 8, 47, 95, 99, 100, 102, 109, 118, 123, 126n4, 131, 169, 192

Fung, H-G., 102
Futures contracts, 8, 18, 28, 43–46, 69, 94–97, 101, 108, 111, 117, 122, 130–135, 139, 142, 145, 156, 159, 171, 174, 176, 178, 180–185, 187, 189, 191

G

Gemmill, G., 93, 99, 100
Generalised method of moments (GMM) approach, 64
  estimator, 58
  parameter, 59
Global financial crisis, 28–30, 195n9
Glosten, L. R., 56
Gonzalo, J., 8
Goodhart, C. A. E., 40
Governmental policy, 3
Goyenko, R.Y., 132
Granger causality analysis, 187–190
Granger, C.W.J., 8, 187–190
Greenhouse gas (GHG) emission, 15, 16, 20, 193
Gregoriou, A., 109, 119
Grossman, S. J., 150, 190

H

Half-effective spread, 68
Hansen, L. P., 58, 59
Harris, L. E., 62
Hasbrouck, J., 3, 8, 149
Hatheway, F. M., 40, 77
He, Y., 40
Hedging, 18, 45, 52, 60, 72, 124, 143
Heflin, F., 65, 84n3

Hendershott, T., 39, 40, 50, 69, 76, 77, 79, 84n6
Hengelbrock, J., 7, 8, 64, 68, 132, 169, 192
Heteroscedastic consistent covariance matrix, 142
Heteroscedasticity and autocorrelation consistent covariance (HAC), 77, 78, 84n3, 140, 142, 143, 178, 185
Hintermann, B., 9, 10n1
Ho, T., 64
Holthausen, R.W., 4, 92, 100
Hrazdil, K., 6, 167, 179, 186
Huang, R. D., 42, 53, 54, 59, 60, 64, 65
Hudson, R., 109, 119

I

Ibikunle, G., 205
Ibrahim, Maher, 186, 194n2
Ice Clear Europe, 97
ICE ECX platform, 148
ICE Futures Europe, 98
ICE platform, 44, 45, 47, 94–96
ICEBLOCK facility, 51
Illiquidity, 3, 5, 167, 181, 183, 186, 187, 195n9, 196n12
Independent software vendors (ISVs), 96
Information absorption, 68–75
Information asymmetry, 3, 30, 42, 51–53, 65–68, 72, 105, 150, 151, 206
Informed traders, 5, 41, 53, 118, 151, 168

Informed trading, 52, 53, 68, 69, 74, 82
Institutional set-up, 94–97
Institutional traders, 4, 52, 96
Institutional trade/trading, 4, 34, 91, 103
Intercontinental Exchange platform (ICE), 44, 94
International Emissions Trading (IET), 16
International Transaction Log (ITL), 16
Interphase banking restrictions, 8
Intertemporal trading, 23–24
Intraday variable, 135
Inventory costs, 59, 61, 151
Inventory holding costs, 61

J

Jarnecic, E., 109
Jarque–Bera normality test, 144
Jiang, C. X., 40
Joint implementation (JI), 16
Joskow, P. L., 18
Joyeux, R., 8, 48

K

Kalaitzoglou, I., 186
Kleidon, A. W., 52
Koch, N., 28
Kollenberg, S., 204
Kossoy, A., 32
Kraus, A., 4, 92
Kurov, A., 82
Kyle, A.S., 3, 165
Kyoto commitment, 19, 23, 24, 130
Kyoto greenhouse gases reduction target, 159
Kyoto Protocol, 15–18

L

Lakonishok, J., 93
Least squares, 65, 76, 78, 178, 185
LEBA carbon index, 140
Lee, C.M.C., 3, 47, 130, 132, 153
Lepone, A., 109
Lesmond, D.A., 132
Liquidity, 1–3, 5–7, 24, 33, 34, 41, 43, 45, 51–54, 56, 58–61, 63, 65, 74, 93, 94, 99, 100, 104, 113, 114, 126n6, 129, 130, 132, 138, 144, 165–194
  improvements, 153–158
  liquidity changes, 154–155
  market resilience measures, 148
  price movements, 148–150
  transaction cost measures, 150–152
  volume-based measures, 147–148
Liquidity-induced trading, 99, 109
Liquidity risks, 6, 129, 167
London Stock Exchange, 93, 98
Long-period returns, 149
Long-term trading volume, 144, 145
Lybek, T., 146

M

Madhavan, A., 6, 40, 53, 54, 83n1, 98, 113, 167
Mansanet-Bataller, M., 9
Markellos, R. N., 9, 33
Market capitalisation, 92

Market confidence, 29, 30, 96, 158–160
Market depth, 97, 102, 109, 148
Market efficiency, 3, 6, 7, 9, 54, 79, 165, 176, 193
Market frictions, 26
Market liquidity, 6, 30, 59, 102, 130, 131, 135, 144, 146, 147, 152, 153, 158, 160, 167, 169, 193, 194
Market maker, 3, 40, 45, 55–59
Market microstructure, 4, 7, 42, 53, 68
Market microstructure models., 151
Market model, 139, 140
Market resilience, 148–150
Market returns, 102, 173, 179
Market stability reserve (MSR), 19, 30, 202–204
Market-efficiency coefficient (MEC), 149
Market-wide liquidity, 147
Markov chain for trade initiation, 57
Maturity contracts, 44, 49, 50, 52, 78, 95, 140
Maximum likelihood, 58, 64
Miclăuș, P. G., 9
Milgrom, P. R., 56
Miller, M. H., 150, 190
Milunovich, G., 8, 9, 48
Mizrach, B., 7, 48, 95, 170
Momentum, 102, 105, 108, 117, 118, 122
Montagnoli, A., 9, 131, 169, 192
Montgomery, W. D., 15
Muckley, C., 9

N

NASDAQ stocks, 6, 40, 47, 50, 77, 167, 186
National Action Plan (NAP), 20, 31
National registry security systems, 31
Nazifi, F., 9
New entrants reserve, 19
New York Stock Exchange (NYSE), 6, 40, 166, 167, 186
Newey and West HAC, 67, 77, 78, 81, 84n3, 142, 143, 178
Noise trading, 96, 169, 191
Nominal order imbalance, 172

O

O'Hara, M., 1, 2
Okereke, C., 205
OLS estimation, 59, 140
OLS estimator, 59, 140
Open interest, 102
Order flow, 55–58
Order flow innovation, 55
Order imbalances, 171–175, 179, 180, 186, 190
Order processing costs, 151
Orthogonality conditions, 64
Otsubo, Y., 7, 48, 95, 170
Over the Counter (OTC) trades, 44, 91, 96, 123, 134, 148

P

Panchapagesan, V., 40
Pardo Tornero, Á., 9
Pástor, L., 6, 167

Patterson, G.A., 102
Payne, R., 54, 130
Pedersen, L.H., 6, 167
Period by period analysis, 74
Permanent price impact, 93, 99–101, 108–110, 112, 114, 117, 118, 122
Permanent price shifts, 149
Person holding account (PHA), 97
Peterson, D. R., 181
Pfleiderer, P., 83n1
Pham, P.K., 132
Platform trading mechanisms, 5
Porter, D. C., 47
Portfolio trading pressure, 59
  extension of model, 63–67
Positive market return coefficient, 109
Predictive regressions, 169, 176–187
Pre-Kyoto commitment trial period, 19
Pre-trading period, 95
Price adjustment process, 2
Price collapse, 10n1
Price concession, 3, 4
Price continuity, 150
Price discovery, 1–3, 5–8, 39–83, 129, 130, 148, 165
  efficiency, 74–79, 84n6
Price formation, 9, 39–41, 54, 149
Price impact, 4, 91–125, 181
  intraday variations in, 110–112
  trade sign and, 105–110
  trade size dependencies on, 113–123
Price movements, 148
Price responsiveness, 25
Price risk, 18
Price transparency, 25, 96

Pricing efficiency, 166, 167, 175, 186, 190
Professional traders, 119
Project emissions permits, 22

Q

Quote driven market, 54
Quoted bid-ask spread, 138, 152, 153
Quoted bid-ask spread ratios, 153
Quote-driven markets, 3, 150, 151
Quoted spread, 55, 62, 132, 158

R

Randomness of returns, 190
Ready, M. J., 3, 47, 132, 153
Recession, 28, 29
Regression, 180, 185
Regular trading hours (RTH), 42, 45, 47, 50, 53, 65, 68, 69, 71, 74, 76, 77, 79
Regulatory arbitrage, 18
Regulatory risk issues, 30
Relative bid-ask spread, 132
Relative spreads, 105, 137, 138, 173–175, 182, 184
Resiliency, 146
Return of Events, 139–142
Revenue generation, 25
Rittler, D., 8
Robustness, 132, 169
Roll, R., 62
Rotfuß, W., 110
Rubin, J. D., 15

Saar, G., 94, 102, 109, 118, 123, 146
Saudi Stock Market (SSM), 109
Schrand, C., 132
Schwartz, R. A., 147, 149
Secondary market, 28, 133, 134
Seguin, P.J., 102
Sell coefficient, 110
Sell trades, 93, 105, 109, 110
Seller-initiated trades, 3, 103, 180
Shaw, K. W., 65
Short-horizon pricing, 169
Short-period returns, 149
Short-run liquidity effects 4
Short-term trading volume changes, 143–144
Sirri, E., 7
Size coefficient, 105
SO_2 emissions, 18
Sovereign debt crisis-induced recession, 202
Speculation, 72
Spot contracts, 8, 52, 133, 193
Spread analyses, 53
Spread Decomposition Model, 54–59
Stambaugh, R.F., 6, 167
Stoll, H.R., 4, 40, 42, 53, 54, 59, 64, 65, 92
Strickland, D., 132
Subrahmanyam, A., 3, 4

Taschini, L., 204
Temporary price impact, 93, 99, 101, 108, 114, 117, 122
Three-way spread decomposition model, 59

Tick rule, 47, 84n4, 125n2
Tick size, 171
Time series estimation, 77
Time series regression model, 142
Tokyo Stock Exchange (TSE), 6
Trade Registration System (TRS), 44, 46
Traded contracts, 191
Traded spread, 173, 182
Trading
  activity measure, 135
  costs, 152
  culture, 118
  emission, 110
  environment, 43–45
  after hours, 39–84
  hour, 102
  motivation for, 51–53
  period, 133
  rules, 156
  system, 43
  tactics, 7
Trading volume, 49, 84n6
  long-term impact of events, 144–145
  short-term impact of events, 142–144
Transaction
  costs, 2, 3, 5, 41, 60, 61, 152
  price, 55, 57, 62
Turnover, 102, 108

Uhrig-Homburg, M., 8, 52
Unbiasedness regressions, 78, 80, 81
Unified regulatory framework, 206
Uninformed traders, 5, 41, 53

U

United Nations Framework Convention on Climate Change (UNFCCC), 17
United States Acid Rain programme, 18

V

Value of assets, 5, 56
Value-weighted index, 139
Van Bommel, J., 84n5
Van Ness, B. F., 64
Variance, 149
  ratios, 190–192
VAT, 30, 31
Vattenfall Energy Trading Netherlands N.V., 133
Vector autoregressive model (VAR), 9, 187, 189
Vector error correction model (VECM), 7
Verrecchia, R., 132

Viswanathan, S., 3, 83n1, 130
Volatility, 49, 50, 101, 113, 119, 126n7, 149

W

Wagner, M., 8, 52
Wald statistics, 187
Walrasian theory, 39
Warner, J.B., 93, 118, 132
Weak form efficiency, 9, 33
Weaver, D. G., 47
WebICE, 96
Weighted price contribution (WPC), 69–73
Weighted price contribution per trade (WPCT), 73–75
Whaley, R. E., 40
White, H., 142
Wilcoxon–Mann–Whitney test, 65, 67, 70, 71, 74, 75
World Bank, 92

GPSR Compliance

The European Union's (EU) General Product Safety Regulation (GPSR) is a set of rules that requires consumer products to be safe and our obligations to ensure this.

If you have any concerns about our products, you can contact us on

ProductSafety@springernature.com

In case Publisher is established outside the EU, the EU authorized representative is:

Springer Nature Customer Service Center GmbH
Europaplatz 3
69115 Heidelberg, Germany

www.ingramcontent.com/pod-product-compliance
Lightning Source LLC
LaVergne TN
LVHW012011060526
838201LV00061B/4271